GOPALAM NEERAJA
S. KRITHIN DATTA
ANAGHA DEVI

Computational Study of MHD Flow with Heat and Mass Transfer in a Porou

GOPALAM NEERAJA
S. KRITHIN DATTA
ANAGHA DEVI

Computational Study of MHD Flow with Heat and Mass Transfer in a Porou

LAP LAMBERT Academic Publishing

Cover image: www.ingimage.com

Publisher:
LAP LAMBERT Academic Publishing
is a trademark of
Dodo Books Indian Ocean Ltd. and OmniScriptum S.R.L publishing group

120 High Road, East Finchley, London, N2 9ED, United Kingdom
Str. Armeneasca 28/1, office 1, Chisinau MD-2012, Republic of Moldova, Europe
Managing Directors: Ieva Konstantinova, Victoria Ursu
info@omniscriptum.com

Printed at: see last page
ISBN: 978-620-9-41213-4

Dr. G. NEERAJA

S. ANAGA DEVI

ANALYSIS OF MAGNETOHYDRODYNAMIC (MHD) FLOW WITH VARIABLE HEAT SOURCE AND CONCENTRATION DIFFUSION

This Book is focused on converts the governing partial differential equations of an MHD boundary layer flow with heat and mass transfer into ordinary differential equations using similarity transformations. The resulting nonlinear ODEs are solved numerically using the fourth-order Runge–Kutta method. The effects of key parameters such as Prandtl number, Schmidt number, buoyancy, and heat generation on velocity, temperature, and concentration profiles are analyzed and presented graphically, highlighting their influence on fluid flow and thermal behaviour.

Authors

S. Anaga Devi
Research Scholar,
Department of Mathematics,
Central University of Andhra Pradesh,
Ananthapuramu.
Dr. G. Neeraja,
Assistant Professor,
Department of Mathematics,
Central University of Andhra Pradesh,
Ananthapuramu.

PREFACE

Chapter 1: This study presents a comprehensive analysis of Burger fluid, a viscoelastic non-Newtonian fluid model that captures both elastic and viscous characteristics. The Burger fluid model is suitable for describing time-dependent deformation, stress relaxation, and creep behavior in complex materials. The influence of magnetic fields on such fluids is examined through the principles of magnetohydrodynamics (MHD). Important dimensionless parameters such as the Prandtl number, Schmidt number, Lewis number, and Eckert number are used to analyze heat and mass transfer phenomena. Buoyancy effects arising from temperature and concentration variations are also incorporated to understand natural convection. The role of internal heat generation on thermal behavior is discussed in detail. Heat and mass transformation mechanisms governing conduction, convection, and diffusion are emphasized. A comparison between Newtonian and non-Newtonian fluids highlights the significance of variable viscosity. The combined effects of viscosity, magnetic fields, and thermal parameters are explored for realistic flow modeling. Overall, this work provides a strong theoretical framework for advanced fluid dynamics and engineering applications.

Chapter 2: This work investigates the nonlinear unsteady boundary layer flow of an incompressible viscous fluid under the combined effects of magnetic field, heat transfer, and mass diffusion. The governing momentum, energy, and concentration equations are transformed into a coupled system of nonlinear ordinary differential equations using similarity transformations. The resulting boundary value problem is solved numerically using the shooting technique in conjunction with the fourth-order Runge–Kutta method, ensuring accurate and stable solutions.

The study contributes to the broader field of magnetohydrodynamic (MHD) and fractional fluid dynamics by building upon earlier investigations that employed similarity methods, finite difference schemes, spectral methods, and fractional calculus to analyze complex flow, thermal, and solutal transport phenomena. Emphasis is placed on understanding the influence of key physical parameters on velocity, temperature, and concentration distributions, with relevance to real-world engineering and industrial applications.

TABLE OF CONTENTS

Chapter Description

List of tables

LIST OF FIGURES

LIST OF SYMBOLS

MHD	Magneto Hydrodynamics
FDM	Finite Difference Method
RK	Runge-Kutta Method
P_r	PrandtlNumber
S_c	Schmidt Number
R_e	Reynolds Number
α	Thermal Diffusivity
β	Buoyancy Parameter
Q_0	Heat Generation Parameter
L_e	Lewis Number
E_c	Eckert Number
G_r	Grashof Number
M	Magnetic Field
σ	Stress Tensor
ε	Strain Rate
η	Dynamic Viscosity
v	Kinematic Viscosity
μ	Dynamic Viscosity
C_p	Specific Heat at Constant Pressure
k	Thermal Conductivity
D	Mass Diffusivity
λ_c	Solutal Buoyancy Parameter
λ_T	Thermal Buoyancy Parameter
T	Local Temperature
T_∞	AmbientTemperature
T_w	Wall Temperature
θ	Dimensionless Temperature
ϕ	Nano particle Volume Fraction

CHAPTER 1

INTRODUCTION

1.1 Burger fluid

Burger fluid is a viscoelastic fluid model that effectively captures both elasticity and viscosity, making it suitable for modeling complex flow behaviors in polymers, biological materials, and geophysical flows. It is developed by combining the Maxwell and Kelvin-Voigt fashions, which collectively seize both brief-time period elasticity and long-time period viscous conduct.

The governing equation for Burger fluid are differential equations integrate relaxation and retardation. Since Newtonian fluid models fail to capture time-dependent deformation, Burger's fluid is often used in studying non-Newtonian fluid dynamics.

This model is applicable in diverse areas, magma flow analysis, biomechanics, and polymer engineering, Additionally, Burger fluid is relevant in magnetohydrodynamics (MHD), thermal conduction, and concentration-driven flows in porous materials. The study of Burger's fluid behavior industrial design, scientific exploration, and development of materials with viscoelastic properties.

$$\sigma + \lambda_1 \frac{d\sigma}{dt} = \eta_1 \left(\varepsilon + \lambda_2 \frac{\partial \varepsilon}{\partial t} \right)$$

where, σ is a stress tensor, ε is a strain rate tensor , η_1 is the viscosity cofficient of the Maxwell element , λ_1, λ_2 is the relaxation and retardation times.

Application of burger fluid are the Geophysics and Earthquake studies this model is applied to analyze mantle dynamics and earthquake faults, aiding in the study of post-seismic deformation. Biomedical applicationIt is helps in modeling blood dynamics with viscoelastic effects and analyze soft tissue biomechanics. Aerospace and Lubrication It is helps analyze the viscoelastic lubricants and dampers, contributing to better shock absorption and vibration isolators. Oil and Gas Industry it plays a crucial role in studying the rheological properties of drilling mud and lubricating fluids. Polymers and soft materials it help's in understanding how polymeric fluids and gels to stress and plays a crucial role in modeling plastic deformation. Asphalt and road engineering it is applied in asphalt pavement design since asphalt shares viscoelastic properties with Burger fluid, aiding in the prediction of rutting and fatigue cracking in roads.

The properties of burger fluid are the Time -Dependent Response is the response of the material changes according to the nature and length of the applied force. Stress Relaxation is the viscosity of the material causes stress to decrease progressively when strain is maintained constant. Viscoelastic Behavior is combines elastic behavior, which can be reversed, with viscous flow, which remains permanent. Non-Newtonian flow is the viscosity of burger fluid is influenced by time and applied stress, unlike Newtonian fluids, which maintain a fixed viscosity. Creep compliance under study stress, it shows immediate elastic deformation before gradually flowing in a viscous manner.

1.2 Magnetohydrodynamics (MHD)

The field of Magnetohydrodynamics (MHD) is a branch of science that focuses on the behavior of electrically conductive fluids when exposed to Magnetic fields. Examples of such fluids encompass plasmas, liquid metals, and saltwater, which interact in distinctive ways with electromagnetic forces. The principle of MHD blend fluid dynamics with electromagnetism, contributing significantly to advancements in physics, engineering, and space science.

MHD is governed by fundamental equations derived from the Navier-stokes equations for fluid mechanics and Maxwell's equations, which govern electromagnetism. The applications of magnetic fields generates Lorentz forces, affecting the movement of the fluid and leading to intricate flow structure. These principles have practical uses in nuclear fusion, space and planetary sciences, geophysical studies, and industrial engineering.

MHD operates under the assumption that the fluid is a strong electrical conductor, permitting magnetic fields to interact with its flow. The magnetic Reynolds number plays a key role in determining if the field remains locked into the fluid or spreads out. When conductivity is high, field lines move along with the fluid, resulting in effect such as plasma instabilities and magnetic reconnection.

Magnetohydrodynamics is utilized in power generation systems, liquid metal cooling for nuclear reactors, and electromagnetic pumps. The ability to manipulate fluid motion through magnetic fields significantly boosts efficiency in industrial fields such as metalworking and fusion plasma containment. MHD is a essential field that connects disciplines like space science, energy production, and materials engineering. Understanding its principles enhances our grasp of fluid motion in material and industrial system, making it a critical element of modern scientific and technological innovation.

1.3 Prandtl number (P_r):

The Prandtl number is crucial parameter that describe the balance between kinetic viscosity and thermal diffusivity in a fluid, influencing heat transfer characteristics. Mathematically, it is expressed as:

$$P_r = \frac{\nu}{\alpha} = \frac{\mu C_p}{k}$$

where, ν is a kinematic Viscosity, α is a thermal diffusivity, μ is a dynamic viscosity, C_p is a specific heat at constant pressure, k is a thermal conductivity, ρ is fluid density.

The Prandtl number plays a crucial role in in defining the relationship between velocity and thermal boundary layers in a fluid. Where (P_r<1), as seen in liquid metals mercury, thermal diffusivity is significantly high, allowing heat to spread quickly. In contrast, high Prandtl number fluids (P_r>1), such as oils and air, experience slower heat diffusion, leading to a dominant velocity boundary layer.

This research explores how changes in the Prandtl number influence temperature distribution and the efficiency of heat transfer. Elevated Prandtl numbers suppress thermal diffusion, intensifying temperature gradients near heated surfaces. Such insights are valuable for optimizing thermal insulation, improving heat exchanger performance, and managing high-temperature industrial applications.

1.4 Schmidt number (S_c):

The Schmidt number is a fundamental dimensionless quantity used to compare momentum and mass diffusion in a fluid. It is determined by the proportion of kinematic viscosity to mass diffusivity.

$$S_c = \frac{\nu}{D}$$

Where $\nu = \frac{\mu}{\rho}$ is the kinematic viscosity (dynamic viscosity μ divided by density ρ), and D is the mass diffusivity of a substance within the fluid. The Schmidt number has significant applications in fields such as heat and mass transfer, aerodynamics, and chemical engineering.

A large Schmidt number ($S_c \gg 1$) indicates that momentum diffusion is more significant than mass diffusion, which is common in liquids due to their high viscosity and slow molecular diffusion. In contrast, a small Schmidt number ($S_c \ll 1$) suggests that mass diffusion is dominant, typically observed in gases.

The Schmidt number is closely linked to other important dimensionless numbers, such as the Prandtl number (P_r) for heat transfer and the Lewis number, which describes the relationship between thermal and mass diffusivities. It plays a vital role in boundary layer theory, aiding in the analysis of mass transfer in turbulent flows, evaporation, and combustion.

In environmental studies, this parameter is essential for analysing pollutant dispersion in fluids, such as air and water. In oceanography, it is used to model the diffusion of dissolved gases like oxygen and carbon dioxide. Similarly, in industrial applications such as cooling towers and chemical reactors, it plays a vital role in optimizing reaction and mixing processes.

Thus, the Schmidt number is a key parameter in engineering and scientific research, offering valuable insights into the relationship between viscosity and mass diffusion.

1.5 Buoyancy parameter ($\boldsymbol{\beta}$):

The buoyancy parameter, a key dimensionless factor in fluid dynamics, describes how buoyancy forces influence flow. It governs natural convection, where fluid motion occurs due to density shifts resulting from thermal or concentration variations.The buoyancy parameter is typically expressed in two forms, depending on whether the buoyancy effect arises from temperature or concentration differences:

Thermal Buoyancy parameter

$$\boldsymbol{\lambda_T = \frac{g\beta_T(T - T_\infty)L}{U^2}}$$

Where, g is the gravitational acceleration, β_T *is a thermal expansion coffecient*, T is a local temperature, $T_\infty$is a ambient temperature, L is a characteristic length, U is a characteristic velocity. This parameter is related to the Richardson number (Ri), where Ri=$\frac{Gr}{Re^2}$ with Gr being theGrashof number and R_e the Reynolds number.

Solutal Buoyancy Parameter (λ_c)

$$\lambda_c = \frac{g\beta C(C - C\infty)L}{U^2}$$

where βc is the solutal expansion coefficient, and C is the concentration.

Buoyancy forces arise due to changes in fluid density caused by temperature or concentration variations, initiating natural convection. When buoyancy is the dominant force, flow patterns develop according to these density changes. In contrast, if inertial effects are stronger, the flow is governed by forced convection, Where mechanical forces like pumps or fans direct movement. The interplay between buoyancy and inertia determines the overall flow behaviour, highlighting the buoyancy parameter's significance in engineering and natural convection studies.

In fluid dynamics, the buoyancy parameter is a crucial dimensionless quantity with diverse applications. It is essential in MHD flows, where it helps evaluate buoyancy-driven behaviour in electrically conducting fluids under the effect of magnetic fields. This parameter plays a significant role in boundary layer theory, particularly in examining thermal and mass transport at the fluid-solid boundary. Moreover, the parameter is significant in atmospheric and oceanic circulation, it provides insight into large-scale fluid motion driven by temperature and salinity gradients, which impact climate and marine currents. Understanding this parameter is crucial in engineering, environmental science, and geophysical studies.

1.6 Heat generation parameter (Q_0):

The heat generation parameter is a dimensionless value that characterizes the effect of internally generated heat in solids and fluids. It is particularly relevant in conduction and convection studies, here heat production results from chemical reactions, nuclear reactions, electrical resistance, or other intrinsic sources. The heat generation parameter is often defined as:

$$Q = \frac{q_0 L^2}{k(T_w - T_\infty)}$$

Where, q_0 is a internal heat generation per unit volume, L is a characteristic length, k is a thermal conductivity of the fluid, T_w wall temperature, $T\infty$ is a ambient temperature.

The heat generation parameter determines how heat is distributed within a fluid or solid medium. When heat generation is positive (Q>0), it contributes to internal heat production, raising the temperature and storing thermal energy. Conversely, negative heat generation (Q<0) leads to heat absorption, producing cooling effects. This parameter governs the behaviour of thermal boundarylayers, influencing the effectiveness of heat transfer in industrial, electronic, and energy-related applications.

The heat generation parameter is essential in many engineering and scientific applications. In magnetohydrodynamic (MHD) flows, it is used to study the effects of heat on electrically conducting fluids under magnetic fields. It is also vital in combustion and chemical reactions, helping to assess heat released by exothermic processes. In electronic cooling, this parameter ensures efficient heat dissipation in microelectronics. Furthermore, it has major applications in nuclear and geothermalenergy, where it helps researchers understand heat generation within reactors and Earth's crust, affecting both power production and geological research.

1.7 Lewis number (L_e):

The Lewis number, symbolized as L_e, is a dimensionless parameter introduced by W.K. Lewis to describe how heat and mass diffuse at different rates in a fluid. It's the Lewis number. defined as the quotient of thermal diffusivity and mass diffusivity., and is expressed as

$$L_e = \frac{\alpha}{D} = \frac{k}{\rho C_p D}$$

Where, α is a thermal diffusivity,D is a mass diffusivity, k is a thermal conductivity, $\rho$is a density, $C_p$is a specific heat capacity at constant pressure.

The (L_e) indicates whether heat or mass transfer is more prominent in a fluid flow by comparing their diffusivities. A $(L_e > 1)$ shows that heat transfer dominates, as thermal diffusion outpaces mass diffusion. When $(L_e < 1)$, mass diffuses faster than heat, making mass transport the leading process. $(L_e = 1)$, both modes progress at equal speeds knowing the Lewis number is fundamental for predicting the outcome of systems involving combined thermal and concentration gradients.

The Lewis number is a key parameter in many scientific and engineering disciplines, particularly where heat and mass transfer interact. In combustion systems, it is used to evaluate flame structure and how heat and species interact during reactions. The number also plays a role in phase change processes, such as evaporation and condensation, where both heat and mass transport are critical. In chemical engineering, it is instrumental in designing and improving units like reactions, dryers, and distillation apparatuses, where achieving the right balance between heat and mass transfer is essential for performance.

1.8 Eckert number (E_c):

The Eckert number (E_c) is a dimensionless figures used in heat transfer and fluid mechanics to link a fluids motion-based energy with the energy difference caused by temperature changes. It plays a crucial role in environments with rapid flow speeds or where internal heating from viscosity is notable, such as in supersonic or compressible flows. It is defined mathematically as

$$E_c = \frac{u^2}{c_p \Delta T}$$

Where, u is a characteristic velocity of the flow, C_p is a specific heat at constant pressure, ΔT is a temperature difference.

The Eckert number reflects the conversion of mechanical energy, stemming from fluid speed, into internal heat as a result of viscous forces. It helps assess how much viscous dissipation contributes to heating within the fluid. A large Eckert number implies that internal energy rises considerably due to velocity-induced heating. Meanwhile, a small value suggests that viscous effects are negligible in the overall energy balance. Engineering rely on this number to decide the importance of viscous heating in a given flow scenario.

The Eckert number has significant utility in thermal systems affected by fluid motion and fractional forces. In aerospace engineering, it helps quantify the thermal impact of high-velocity travel, such as during re-entry or supersonic flight. For compressible flow conditions, the number highlights where internal heating due to viscous dissipation must be considered. It's also crucial in lubrication science, where it helps predict temperature changes within narrow fluid films caused by friction.

1.9 Heat Transformation:

The process of heat transformation involves transferring or converting heat energy from one system to another. This can happen via conduction (through solids), convection (through fluids), or radiation (without a medium). This process is essential in thermodynamics, playing a crucial role in power generation, cooling technologies, and heat exchanger design.

Conduction transfers heat through solids, convection moves it within fluids, and radiation enables heat transfer through electromagnetic waves, such as sunlight. The study of heat transformation is important in engineering, climatecontrol,and energy-efficient technologies.

1.10 Mass Transformation:

Mass transformation describes the transfer or conversion of mass between different phases, states, orsystems. It is an important aspect of various natural and technological processes, such as chemical reactions, phase changes, and fluid transport. In chemical processes, mass transformation occurs when substances undergo reactions that modify their structure. Phase changes, such as and melting, boiling,condensation, involve the movement of mass between solid, liquid, and gaseous states. In fluiddynamics, mass transfer happens through diffusion, convection, and advection. Biological functions, including metabolism and nutrient exchange, depend on mass transformation. It is also essential in combustion, material synthesis, and industrial production. This concept is widely applied in engineering, environmental science, and material research.

1.11 Newtonian and Non-Newtonian Fluid:

Newtonian fluids are characterized by a constant viscosity, following Newton's law of viscosity, where shear stress is proportional to shear rate. Their viscosity remains unchanged regardless of the force applied or the flow conditions. Examples of Newtonian fluids include water, air, and manypure liquids. These fluids follow a consistent shear stress-strain rate behaviour. They play a significant role in fluid mechanics and industrial processes.

Non-Newtonian fluids are characterized by a viscosity that changes with applied force, meaning they do not obey Newton's law of viscosity. Unlike Newtonian fluids, these materials have a nonlinear stress-strain relationship. Examples include blood, ketchup, toothpaste, and paint, which display various flow behaviors. Some become less viscous when stirred (shear-thinning), while others become thicker (shear-thickening) or display viscoelasticity. These fluids are crucial in biomechanics, industrial coatings, and food engineering.

1.12 Viscosity:

Viscosity defines a fluid's internal resistance to movement and deformation. It reflects the internal friction between adjacent layers as they move. Fluids like honey, which flow slowly, have high viscosity, while water has low viscosity and flows easily. This behaviour is governed by the strength of molecular forces within the fluid. Viscosity is a fundamental concept in fluid flow, affecting speed, drag, and energy consumption. It is either dynamic or kinematic, depending on density inclusion.

CHAPTER-2

ANALYSIS OF MAGNETOHYDRODYNAMIC (MHD) FLOW WITH VARIABLE HEAT SOURCE AND CONCENTRATION DIFFUSION

2.1 Introduction

This work investigates the nonlinear convective flow of an incompressible viscous fluid subjected to magnetic, thermal, and solutal effects. The governing partial differential equations which model momentum, energy, and concentration transport in a two-dimensional unsteady boundary layer flow, are transformed into a system of ordinary differential equations using a similarity transformation technique. The similarity variables reduce the complex coupled PDE system into a nonlinear system of ODEs involving the dimensionless concentration, temperature and velocity profiles.

To solve the resulting boundary value problem, a numerical approach combining the shooting method and thefourth-order Runge-Kutta method (RK4) is used. The shooting method is used to iteratively determine the missing initial conditions that satisfy the far-field boundary conditions. The RK4 method provides high-accuracy numerical integration over the similarity domain.

The investigation deals with MHD flow incorporating thermal and solutal effects. The flow-regulating system of PDEs is transformed into a set of ODEs using similarity transformation. The Runge-Kutta method is used to solve these equations numerically, making it easier to analyze the effect of physical parameters on flow velocity, heat distribution, and species concentration in depth. Yuehua Jiang [1] Using the fractional Burgers model, this study examines the effects of chemical reactions, radiation, and periodic boundary conditions on the flow behaviour, heat transfer, and mass transfer of non-Newtonian fluids in porous media. Yasir Mahsud [2] Fractional constitutive equations and Laplace transform techniques are used to derive solutions to problems involving unsteady, nonlinear boundary layer convection flows and heat transfer of a fractional Maxwell fluid near a vertical plate with constant thermal flux in this study. Jinghua Zhang [3]Using the corrected weighted shifted Grünwald difference scheme and the Legendre spectral method, this study proposes an analytical solution and a unified numerical framework for the multi-term time-fractional Burgers' fluid model (MT-TFBFM). This ensures stability, accuracy, and effectiveness.Zefei Long [4] Using the finite difference method, the study develops a modified unsteady Marangoni convection model with distributed fractional derivatives, investigates thermal and flow dynamics, and compares the accuracy of the numerical scheme to exact solutions. Yang Weidong [5] This research analyzes the fractional boundary layer flow and heat transfer behavior over a non-uniform stretching sheet using a double fractional Maxwell model, highlighting the finite difference method's effectiveness in resolving the role of fractional parameters. Shuo yang [6]The paper presents a distributed order time-fractional model for analyzing unsteady natural convection flow and thermal transport of viscoelastic fluids under magnetic field effects, using numerical techniques to explore and compare parameter impacts. M.A. Imran [7] By employing the Caputo-Fabrizio fractional derivative, this study provides an exact analysis of natural convection flow in a non-Newtonian Jeffrey fluid, taking into account a variety of physical effects and demonstrating how fractional order increases velocity and decreases skin friction. Xiaoqing Chi [8] Using the finite difference spectral decomposition method to solve the nonlinear coupled model and examine how key parameters affect the flow dynamics and temperature distribution, the study investigates unsteady space fractional MHD free convective flow and heat transfer with Hall effects. Shan Shi [9] A rapid numerical scheme based on the Legendre spectral method and the sum-of-exponentials technique is presented in this paper to investigate fractional MHD flow of a generalized second-grade fluid in a porous medium using Hall current. Its effectiveness is demonstrated through theoretical analysis and numerical simulations. V. Nagendramma [10] Using slip boundary conditions, thermal radiation, and viscous dissipation, a steady-state investigation of an incompressible Maxwell nanofluid's steady flow, heat exchange, and mass diffusion over a stretching sheet is conducted. The governing equation are converted into nonlinear ODE through suitable similarity transformation. A suitable transformation leads to nonlinear equations, which are solved for convergent solutions. The inclusion of buoyancy forces, heat absorption/generation, viscous dissipation, and concentration effects makes the study more comprehensive for real-world applications.

2.2 Formulation of the problem:

In this study, we analyse unsteady MHD flow of a Burgers fluid over a stretching surface under both linear and nonlinear conditions. The solute concentration inside the boundary layer (C_w , C_∞)varies from the ambient fluid, while the temperature (T_w , T_∞) is assumed similar across the boundary. The stretching velocity is defined as $u_w = \frac{ax}{1-ct}$, where a is the rate of stretching and ccc denotes the positive unsteadiness parameter. The X-axis lies along the stretching sheet, and the Y-axis is perpendicular. Viscous dissipation and heat generation/absorption are also integrated into the governing equations.

$$\frac{\partial u}{\partial x} + \frac{\partial v}{\partial y} = 0 \tag{2.1}$$

$$\frac{\partial u}{\partial t} + u\frac{\partial u}{\partial x} + v\frac{\partial u}{\partial y} + \alpha_1\left(2uv\frac{\partial^2 u}{\partial x\partial y} + u^2\frac{\partial^2 u}{\partial x^2} + v^2\frac{\partial^2 v}{\partial y^2}\right) + \alpha_2\left[u^2\left(2\frac{\partial^2 u}{\partial x\partial y} + \frac{\partial^2 u}{\partial x^2}\frac{\partial u}{\partial x} - \frac{\partial^2 v}{\partial x^2}\frac{\partial u}{\partial y}\right) + u^3\frac{\partial^3 u}{\partial x^3} + v^3\frac{\partial^3 u}{\partial y^3} + 3uv\left(u\frac{\partial^3 u}{\partial x^2\partial y} + v\frac{\partial^3 u}{\partial x\partial y^2}\right) + 3v^2\left(\frac{\partial^2 u}{\partial x\partial y}\frac{\partial u}{\partial y} + \frac{\partial^2 u}{\partial y^2}\frac{\partial u}{\partial y}\right)\right] - \frac{\sigma\beta_0^2}{\rho} = v\left[\alpha_3\left(u\frac{\partial^3 u}{\partial x\partial y} + v\frac{\partial^3 u}{\partial y^3} - \frac{\partial u}{\partial x}\frac{\partial^2 u}{\partial y^2} + \frac{\partial u}{\partial y}\frac{\partial^2 v}{\partial y^2}\right)\right] + g\beta_T(T_0 - T_\infty) + g\beta_T(T_0 - T_\infty)^2 + g\beta_C(C_0 - C_\infty) + g\beta_C(C_0 - C_\infty)^2$$

(2.2)

$$\frac{\partial T}{\partial t} + u\frac{\partial T}{\partial x} + v\frac{\partial T}{\partial y} = \kappa\frac{\partial^2 T}{\partial y^2} + \tau\left[D_\beta\frac{\partial C}{\partial y}\frac{\partial T}{\partial y} + \frac{D_T}{T_\infty}\left(\frac{\partial T}{\partial y}\right)^2\right] + q^m + \frac{1}{\rho C_p}\left[\mu\left(\frac{\partial u}{\partial y}\right)^2\right] \tag{2.3}$$

$$\frac{\partial C}{\partial t} + u\frac{\partial C}{\partial x} + v\frac{\partial C}{\partial y} = \frac{D_T}{T_\infty}\frac{\partial^2 T}{\partial y^2} + D_B\frac{\partial^2 C}{\partial y^2} \tag{2.4}$$

The governing PDEs for this flow must satisfy the respective boundary conditions are

$$At\, y \rightarrow 0 \;\; then\; u = u_w,\; v = 0,\; T_0 = T_w,\; C_0 = C_w$$

$$At\, y \rightarrow \infty \;\; then\; u = 0, T_0 \rightarrow T_\infty,\; C_0 \rightarrow C_\infty \quad (2.5)$$

By introducing dimensionless stream function, the mathematical complexity of the problem is reduced, achieved through a similarity transformation that converts the PDEs into ODEs.

$u = \left(\frac{ax}{1-ct}\right) f'(\eta)$, $\eta = y\left(\frac{a}{v(1-ct)}\right)^{\frac{1}{2}}$, $v = -\left(\frac{av}{1-ct}\right)^{\frac{1}{2}} f(\eta)$

$T = T_\infty + (T_w - T_\infty)\theta(\eta)$, $C = (C_w - C_\infty)\emptyset(\eta)$ **(2.6)**

where η is the similarity transformation, $f(\eta), \theta(\eta), and\ \phi(\eta)$ are the dimensionless stream function, temperature and concentration functions differentiate with respect to η.

The governing equations are converts PDE into ODE

$$f''' + \frac{1}{2} f f'' - \beta\theta - \beta_C\phi = 0 \qquad (2.7)$$

$$\theta'' + \Pr\left(\frac{1}{2} f\theta' + Q_0 + Q_1\theta\right) = 0 \qquad (2.8)$$

$$\phi'' + Sc\left(\frac{1}{2} f\phi'\right) = 0 \qquad (2.9)$$

With boundary conditions in ODE form

At $\eta = 0\ f(0) = 0$, $f'(0) = 0$, $\theta(0) = 1$, $\phi(0) = 1$

At $\eta = \infty\ f'(\infty) = 0$, $\theta(\infty) = 0$, $\phi(\infty) = 0$ **(2.10)**

The governing equations convert into Runge-Kutta method (RK4)

We define: $f_1 = f$, $f_2 = f'$, $f_3 = f'''$

$\theta_1 = \theta$, $\theta_2 = \theta'$

$$\phi_1 = \phi \ , \ \phi_2 = \phi' \qquad (2.11)$$

$$f_1' = f_2 \qquad (2.12)$$

$$f_2' = f_3 \qquad (2.13)$$

$$f_3^1 = -\frac{1}{2} f_1 f_2 + \beta\theta_1 + \beta_C\phi_1 \qquad (2.14)$$

$$\theta_1' = \theta_2 \qquad (2.15)$$

$$\theta_2' = -\Pr\left(\frac{1}{2} f_1\theta_2 + Q_0 + Q_1\theta_1\right) \qquad (2.16)$$

$$\phi_1' = \phi_2 \qquad (2.17)$$

$$\phi_2' = -Sc\left(\frac{1}{2}f_1\phi_2\right) \quad (2.18)$$

Where the boundary conditions are:

$$\eta = 0\ f_1(0) = 0,\ f_2(0) = 1,\ \theta_1(0) = 1,\ \phi_1(0) = 1$$

$$\eta = \infty f_2(\infty) = 0,\ \theta_1(\infty) = 0, \phi_1(\infty) = 0 \quad (2.19)$$

2.3 Results:

The velocity profiles corresponding to β values of 0.5, 1.0, and 1.5 are depicted in Figure 2.1 to 2.3. As β increases, notable changes occur in both the peak velocity and boundary layer thickness. At (Figure 2.1) $\beta = 0.5$ the velocity rise more gradual, reflecting a weaker convective effect. Similarly, (Figure 2.2) at $\beta = 1.0$ the velocity reaches a greater maximum and the boundary layer becomes thinner. Figure 2.3 β further increased toillustrate the effects of the magnetic field on velocity and radiation on temperature, respectively. the velocity gradient near the boundary becomes steeper, implying stronger momentum transport. These results suggest that β intensifies fluid motion and refines velocity gradients, shaping overall flow dynamics. Lastly. (Figure 2.4) in the magnetic field suppresses velocity by exerting the Lorentz force, which opposes fluid motion.

The temperature profiles for varying Prandtl numbers (P_r=0.71, 1.00, and 7.00) are depicted in Figure 2.5 to 2.8. As pr increases, the thermal boundary layer thickness decreases noticeably. At pr=0.71 (Figure 2.5), the temperature spreads over a wider region, indicating greater thermal diffusivity and efficient heat transfer. In (Figure 2.6) P_r=1.00 the profile steepens, indicating a moderate reduction in boundary layer thickness. For P_r=7.00 (Figure 2.7), the temperature gradient near the boundary becomes more pronounced, suggesting reduced thermal diffusivity and a more concentrated heat transfer zone. Illustrate the (Figure 2.8) influence of the Eckert number (E_c) on temperature profiles. These observations highlight that increasing S_c results in a thinner thermal boundary layer and enhanced temperature gradients.

The effect of the (S_c) and permeability on concentration profiles are depicted in Figure 2.9 to 2.10. As shown in Figure 2.9, a higher S_c leads to a steeper concentration gradient and thinner concentration boundary layer, suggesting a reduction in mass diffusivity. Increased S_C restricts molecular diffusion, resulting in a more abrupt concentration drop away from the surface. On the other hand, Figure 2.10 demonstrates the effect of permeability, where increased permeability enhances mass transfer by allowing fluid to flow more easily through the porous medium. As permeability increases, the concentration profile becomes broader, reducing sharp concentration gradients. These results emphasize the opposing effects of S_c and permeability: higher S_c suppresses diffusion, whereas increased permeability enhances it.

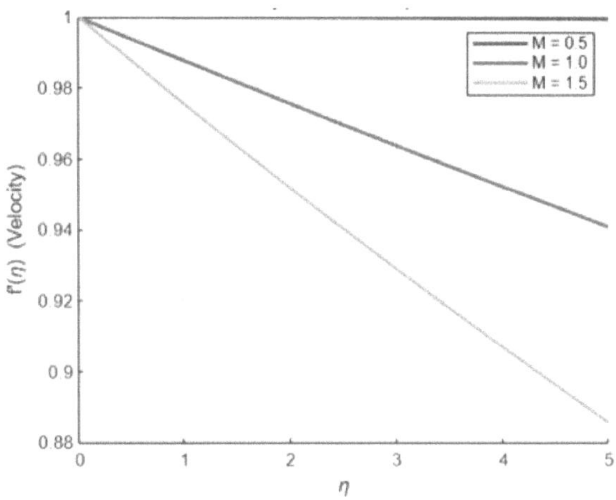

Fig.2.1 Velocity profile for $\beta = 0.5$

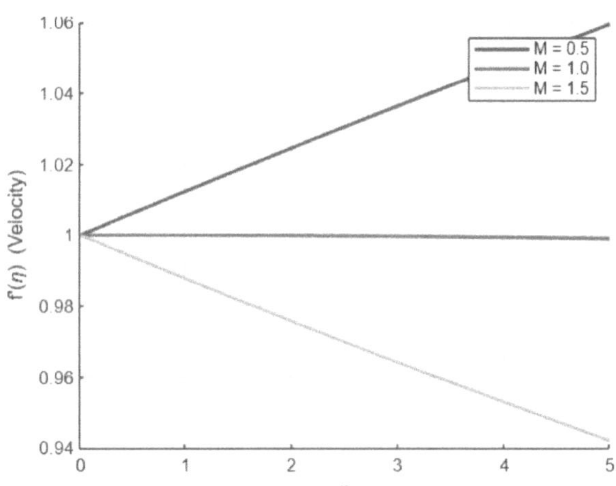

Fig.2.2 Velocity profile for $\beta = 1.0$

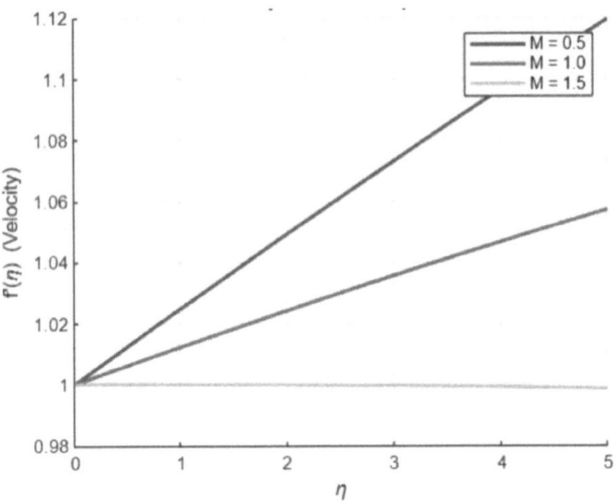

Fig.2.3 Velocity profile for β = 1.5

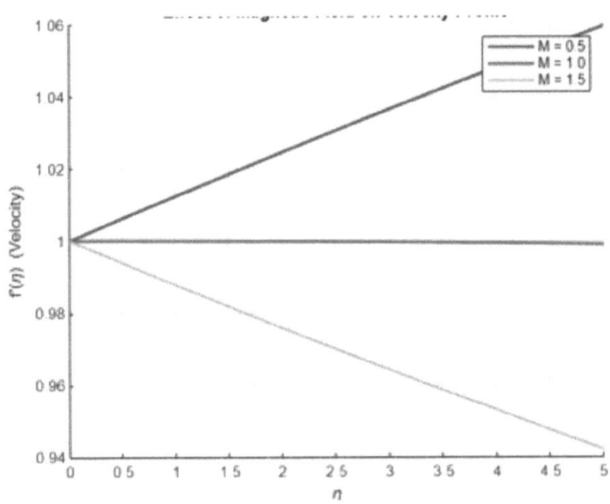

Fig.2.4 Effect of Magnetic Field on velocity profile

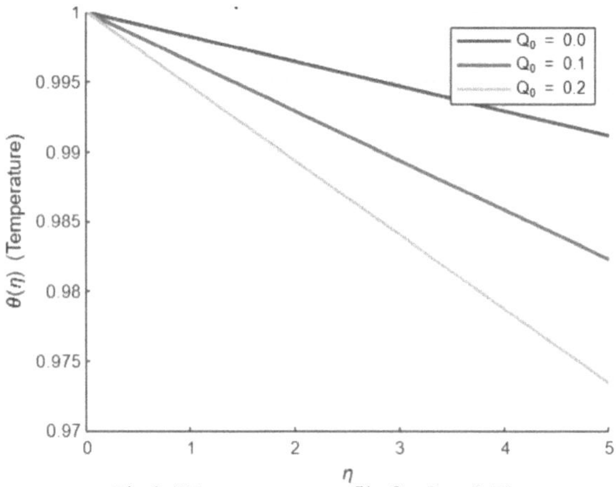

Fig.2.5 Temperature profile for P_r = 0.71

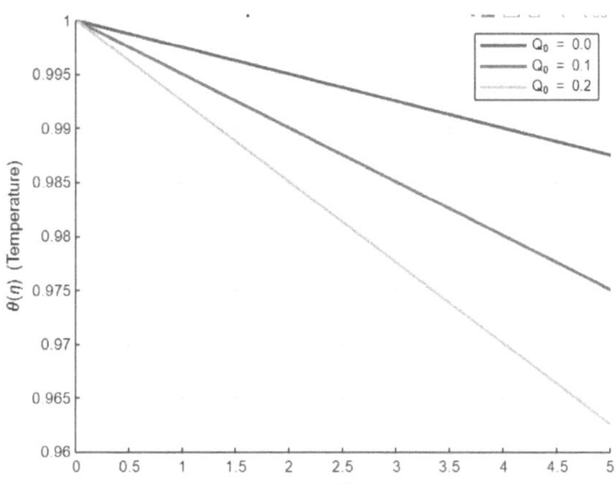

Fig.2.6 Temperature profile for P_r = 1.00

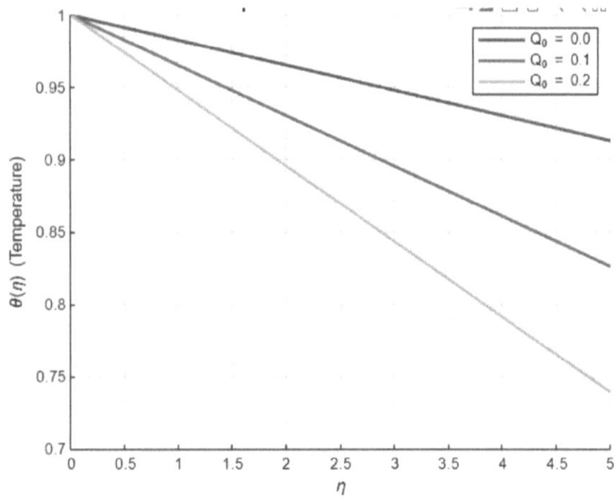

Fig.2.7 Temperature profile for $P_r = 7.00$

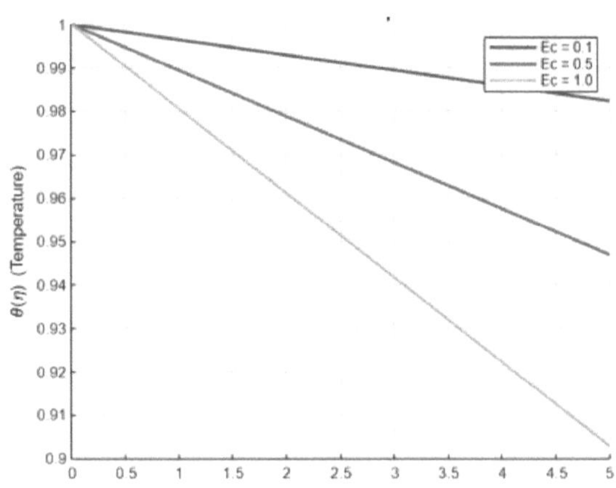

Fig.2.8 Effect of radiation on Temperature profile

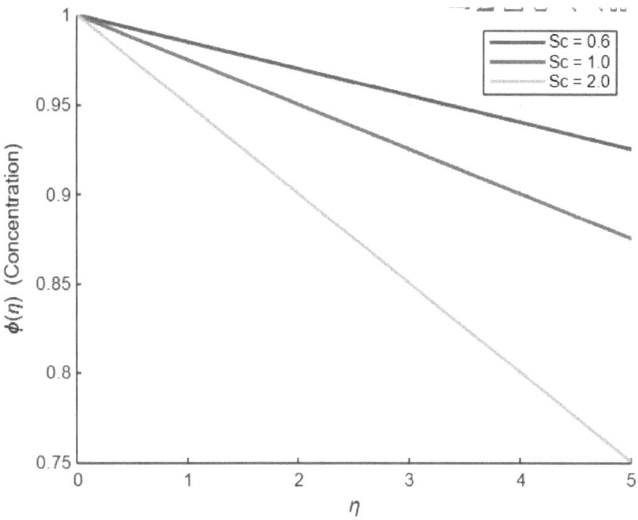

Fig.2.9 Effect of Schmidt number on concentration

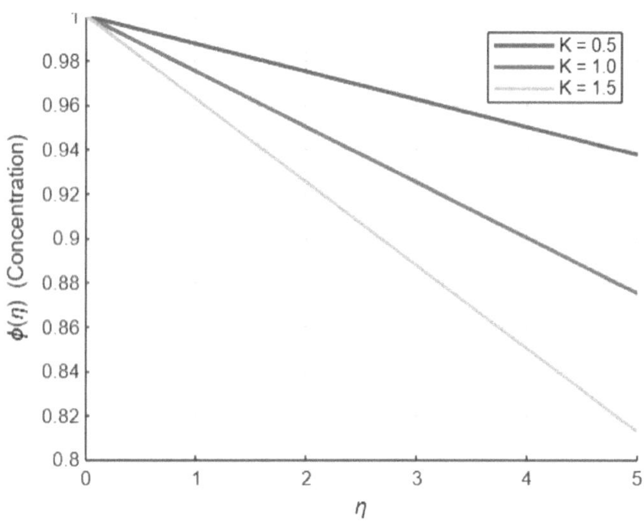

Fig.2.10 Effect of permeability on concentration

For table of velocity, temperature, and concentration profiles at selected values of η. The tables will show:

1. Velocity $f'(\eta)\ for$ different β
2. Temperature $\theta(\eta)\ for$ different P_r
3. Concentration $\phi(\eta)\ for$ different S_c
4. Effect of heat generation Q_0 on temperature

Table 2.1 Velocity $f'(\eta)$ for different β, Temperature $\theta(\eta)$ for different P_r and Concentration $\phi(\eta)$ for different S_c values.

η	$f'(\eta)$	$\theta(\eta)$	$\phi(\eta)$
0.0	1.0000	1.0000	1.0000
0.5	0.8752	0.9021	0.9435
1.0	0.7623	0.8145	0.8912
1.5	0.6591	0.7362	0.8423
2.0	0.5643	0.6662	0.7961
2.5	0.4767	0.6038	0.7525
3.0	0.3956	0.5482	0.7112
3.5	0.3204	0.4986	0.6721
4.0	0.2507	0.4546	0.6350
4.5	0.1860	0.4156	0.6000
5.0	0.1261	0.3811	0.5667

1. Velocity profile vs Buoyancy parameter (β):

The velocity profile f′(η) significantly depends on the buoyancy parameter β, which represents the effect of thermal buoyancy forces in the fluid. Below is an analysis of how β influences velocity in MHD flow.

2. Temperature profile $\theta(\eta)$ vs Prandtl number (P_r):

The (P_r) significantly influences the temperature profile in fluid flows by governing the relationship between momentum diffusivity and thermal diffusivity. A higher Prandtl number implies that heat diffusion is slower compared to momentum diffusion, leading to steeper temperature gradients near heated surfaces. Conversely, a lower Prandtl number indicates faster heat diffusion, resulting in a more uniform temperature distribution.

3. Concentration Profile $\phi(\eta)$ vs S_c (Schmidt Number):

The concentration profile ϕ(η) varies significantly with the Schmidt number (S_c), which governs the relative effects of momentum and mass diffusion in a fluid system. Lower S_c Higher mass diffusivity relative to viscosity, more uniform concentration distribution, Broader concentration boundary layers. Higher S_c Lower mass diffusivity, meaning species diffusion is slower, Steeper concentration gradients near the surface, thinner mass boundary layers.

2.4 CONCLUSION

This research successfully explores the behaviour of numerical scheme proved to be robust and accurate, effectively capturing the velocity, temperature, and concentration fields under varying physical parameters such as the Prandtl number, Schmidt number, and heat generation/absorption rates. The results demonstrate that thermal and solutal buoyancy significantly influence the momentum and thermal boundary layers. Furthermore, internal heat generation tends to thicken the temperature profile, while higher Schmidt numbers reduce the concentration boundary layer. Overall, the numerical approach used in this work provides a reliable framework for analysing similar transport phenomena in engineering and applied science domains.

References:

1. Y. Jiang, H. Sun, Y. Bai, Y. Zhang, MHD flow, radiation heat and mass transfer of fractional Burger's fluid in porous medium with chemical reaction comput. Math. Appl. 115 (2022) 68-79
2. Y. Mahsud, N.A. Shah, D. Vieru, Natural convection flows and heat transfer with exponential memory of a Maxwell fluid with damped shear stress, Comput. Math. Appl. 76 (2018)
3. J. Zhang, F. Liu, Z. Lin, V. Anh, Analytical and numerical solutions of a multi-term time-fractional Burgers' fluid model, Appl. Math. Comput. 356 (2019) 1–12.
4. Z. Long, L. Liu, S. Yang, L. Feng, L. Zheng, Analysis of Marangoni boundary layer flow and heat transfer with novel constitution relationships, Int. Commun. Heat Mass Transf. 127 (2021) 105523
5. Y. Weidong, C. Xuehui, Z. Xinru, Z. Liancun, L. Fawang Flow and heat transfer of double fractional Maxwell fluids over a stretching sheet with variables thickness. Appl. Math. Model. 80 (2020) 204-216
6. S. Yang, L. Liu, Z. Long, L. Feng, Unsteady natural convection boundary layer flow and heat transfer past a vertical flat plate with novel constitution models, Appl. Math. Lett. 120 (2021) 107335
7. M.A. Imran, F. Miraj, I. Khan, I. Tlili, MHD fractional Jeffrey's fluid flow in the presence of thermo diffusion, thermal radiation effects with first order chemical reaction and uniform heat flux, Results Phys. 10 (2018) 10–17.
8. X. Chi, H. Zhang, Numerical study for the unsteady space fractional magnetohydrodynamic free convective flow and heat transfer with Hall effects, Appl. Math. Lett. 120 (2021) 107312.
9. S. Shi, X. Jiang, H. Zhang, A fast method and convergence analysis for the MHD flow model generated second-grade fluid. Comp. Math. Appl. 171 (2024) 175-187.
10. V. Nagendramma, R.V.M.S.S. Kiran kumar, P. Durga prasad, A. Leelaratnam, V.K. Varma, Multiple slips and Thermophoresis Effect of Maxwell Nanofluid over a permeable stretching surface in the presence of Radiation and Dissipation. Appl. Math. 5 (2016) 1-9.
11. G.K. Ramesh, B.J. Gireesha, Rama Subba Reddy Gorla, (Boundary layer flow past a stretching sheet with fluid-particle suspension and convective boundary condition. 51 (2016) 1061-1066.

12. S. Yang, L. Liu, Z. Long, L. Feng, Unsteady natural convection boundary layer flow and heat transfer past a vertical flat plate with novel constitution models, Appl. Math. Lett. 120 (2021) 107335.
13. W. Yang, X. Chen, X. Zhang, L. Zheng, F. Liu, Flow and heat transfer of double fractional Maxwell fluids over a stretching sheet with variable thickness, Appl. Math. Model. 80 (2020) 204–216.
14. Iskandar Waini, Anuar Ishak, Ioan Pop, Unsteady flow and heat transfer past a stretching/shrinking sheet in a hybrid nanofluid, International Journal of Heat and Mass transfer. 136 (2019) 288-297.
15. M. Tamoor, M. Waqas, M. Ijaz Khan, Magnetohydrodynamic flow of casson fluid over a stretching cylinder. Results in Physics. 7. (2017) 498-502.

Dr. G. NEERAJA

S. KRITHIN DATTA

Computational Study of MHD Flow with Heat and Mass Transfer in a Porous Medium

This Book is focused on a three-dimensional magnetohydrodynamic (MHD) boundary layer flow under a transverse magnetic field with heat and mass transfer effects. The governing equations are transformed into ordinary differential equations using similarity transformations and solved numerically by the finite difference method. The effects of physical parameters on velocity, temperature, and concentration profiles are presented graphically.

Authors

S. Krithin Datta
Research Scholar,
Department of Mathematics,
Central University of Andhra Pradesh,
Ananthapuramu.
Dr. G. Neeraja,
Assistant Professor,
Department of Mathematics,
Central University of Andhra Pradesh,
Ananthapuramu.

PREFACE

Chapter 1: In the first chapter, we discussed the basic introduction and someapplications for the study of Maxwell fluids is essential for understanding viscoelastic flow behavior encountered in many engineering and scientific applications. Unlike Newtonian fluids, Maxwell fluids exhibit both elastic and viscous characteristics, allowing them to model memory and stress relaxation effects accurately. When such fluids interact with magnetic fields, magnetohydrodynamic (MHD) effects significantly influence flow, heat, and mass transfer mechanisms. Dimensionless parameters such as the Prandtl, Schmidt, Soret, and Dufour numbers play a crucial role in governing thermal and concentration distributions. Thermophoresis and Brownian motion further affect particle transport in complex fluid systems. These coupled physical phenomena arise in polymer processing, biomedical engineering, geophysics, and energy systems. Due to the nonlinear nature of the governing equations, analytical solutions are often limited. Hence, numerical techniques such as the finite difference and finite element methods are employed for accurate analysis. This study aims to provide a comprehensive numerical investigation of MHD Maxwell fluid flow with heat and mass transfer effects.

Chapter 2: Magnetohydrodynamic (MHD) flow analysis is of great importance due to its wide range of applications in engineering and applied sciences. The interaction between magnetic fields and electrically conducting fluids significantly alters fluid motion, heat transfer, and mass diffusion characteristics. In recent years, nanofluids and non-Newtonian fluid models, such as Maxwell fluids, have gained attention for their enhanced thermal performance. The present study focuses on a three-dimensional MHD boundary layer flow of an upper-convected Maxwell nanofluid over a bilinear stretching surface. The combined effects of magnetic field, thermal radiation, Brownian motion, thermophoresis, and chemical reactions are incorporated into the model. Similarity transformations are employed to reduce the governing partial differential equations into ordinary differential equations. The resulting system is solved numerically using the finite difference method. The influence of various physical parameters on velocity, temperature, and concentration fields is examined. The outcomes of this study are relevant to advanced cooling systems, heat exchangers, and MHD-based industrial applications.

TABLE OF CONTENTS

List of tables

Table No **Description**

LIST OF FIGURES

FIGURE No **Description**

LIST OF SYMBOLS

MHD	MagnetoHydrodynamics
FDM	Finite Difference Method
FEM	Finite Element Method
FVM	Finite Volume Method
PDE	Partial Differential Equation
ODE	Ordinary Differential Equation
HAM	Homotopy Analysis Method
S_c	Schmidt Number
α	Thermal Diffusivity
D_T	Dufour Number
S_r	Soret Number
D_B	Brownian Diffusion
K_T	Thermophoretic Parameter
L_e	Lewis Number
M	Magnetic Field
v	Kinematic Viscosity
μ	Dynamic Viscosity
C_p	Specific Heat at Constant Pressure
k	Thermal Conductivity
D	Mass Diffusivity
θ	Dimensionless Temperature
ϕ	Nano particle Volume Fraction
H_a	Hartmann Number
D_a	Darcy Number
T	Local Temperature
T_w	Wall Temperatur
T_∞	Ambient Temperature

CHAPTER 1
INTRODUCTION

1.1 Maxwell fluid

Maxwell fluids are a type of non-Newtonian fluid that simultaneously display both thick consistency and elastic behavior making it is important model in the viscoelastic fluid mechanics. It was first introduced by James Clerk Maxwell in 1867 to explain how certain materials behave as both like solid and liquid characteristics when subjected to stress. The Maxwell model is constructed by using Hookean elastic spring and a Newtonian viscous component connected in series, which helps to capture the viscoelastic effects effectively.

Newtonian fluids follow a linear stress-strain relationship, Maxwell fluids obey a constitutive equation that integrates a relaxation time parameter.How long the influence of past strain persists in a fluid before it transitions to Newtonian manner. The maxwell fluid is mathematically governed by the equation:

$$\frac{d\tau}{dx} + \lambda \frac{d}{dt}(\Delta v + (\Delta v)^T) = 2\eta \Delta v$$

Where $\boldsymbol{\tau}$is the stress tensor, $\boldsymbol{\lambda}$denotes the response time, $\boldsymbol{\eta}$ represents the viscous coefficient, and $\boldsymbol{v}$ refers fluid velocity.

Maxwell fluids present viscoelastic responses under stress, and can flow like a fluid but also recover like an elastic solid.One of their key features is stress relaxation, where their stress gradually diminishes when strain remains constant.The model assumes a single relaxation time, making it particularly useful for describing polymeric and other non-Newtonian fluids and polymeric materials.Additionally, depending on modifications, Maxwell fluids may exhibit shear thinning or thickening behaviour. Unlike Newtonian fluids, they do not deform instantaneously when exposed to external forces due to their elastic component.Their behaviour is also influenced by the Deborah number, which helps determine how their behaviour changes over different time scales. Another important feature is their memory effect, meaning their deformation depends on both past and present stress conditions, making them highly time-dependent.

The Maxwell fluid model, which characterizes viscoelastic behaviour, finds application in several fields.It is widely used in polymer processing, helping engineers study the flow of polymer melts and solutions in molding and extrusion. The model is also crucial in biological fluid mechanics, where it helps in understanding the viscoelastic nature of blood and synovial fluid dynamics.The oil and gas sector leverages this model to analyse the transport of crude Crude oil flow and drilling mud behaviour in porous geological formations.Similarly, in the food industry, it aids in the study of dough elasticity and the viscosity of dairy products.The model is also used in geophysics to simulate magma and lava flow.while aerospace engineers use it to improve non-Newtonian lubricant performance in spacecraft.It also contributes to biomedical applications, particularly in drug delivery systems that require precise control over fluid dynamics.The textile industry benefits from its applications in Fiber spinning and fabric coating.Finally, MHD applications rely on Maxwell fluid properties to study electrically conductive viscoelastic fluids in astrophysics and industrial processes.

Since Maxwell fluids exhibit nonlinear behaviour, computational methods such as FEM and FDM are required for accurate analysis. These techniques are essential for solving nonlinear partial differential equations, particularly in complex flow of geometrical configurations or perturbation analysis.

1.2 Magnetohydrodynamics (MHD)

Magnetohydrodynamics (MHD) is a branch of physics that examines the motion with respect to electrically conductive fluids under magnetic field effects. It combines fluid mechanics and electromagnetism to describe the behaviour of plasmas, molten metals, and ionized fluids. The primary equations of MHD originate from the Navier-Stokes and Maxwell's equations, with Ohm's law linking the electric and magnetic interactions. The movement of the fluid within a magnetic field generates Lorentz forces, which significantly alter the system's dynamics.

MHD is integral to several technological and scientific domains, including fusion energy, astrophysics, liquid metal applications in nuclear reactors, and electromagnetic propulsion. In space science, MHD governs solar wind interactions, accretion disks, and stellar magnetic fields. In power generation, MHD generators directly convert kinetic energy into electrical energy, eliminating the need for moving parts.

MHD flow studies encompass channel flows, boundary layers, and convective systems. Magnetic fields influence velocity, heat transfer, and species concentration, which is particularly relevant in plasma control and high-temperature systems

As a crucial research area, MHD continues to provide insights into energy systems, advanced cooling technologies, and the behaviour of electrically conductive fluids in both laboratory and astrophysical settings.

1.3 Prandtl number (P_r):

The Prandtl number, commonly denoted as $\boldsymbol{P_r}$, characterizes the relative thickness of the momentum and thermal boundary layers by comparing kinematic viscosity to thermal diffusivity.

$$\boldsymbol{P_r = \frac{v}{\alpha} = \frac{\mu C_p}{k}}$$

where v stands for kinematic viscosity, α represents thermal diffusivity, μ is the shear viscosity, C_p denote pressure-specific heat, and k represent thermal transmission rate.

When P_r is low, such as in liquid metals ($P_r \ll 1$), thermal diffusivity is significantly higher than momentum diffusivity, resulting in a thin thermal boundary layer. Conversely, when P_r is high, as in oils ($P_r \gg 1$), momentum diffusivity dominates, this causes the thermal boundary layer to grow considerably thicker than the velocity boundary layer.

Gases like air have $P_r \approx 0.7$, while water has $P_r \approx 7$. High-P_r fluids have lower heat transfer efficiency. The Prandtl number is crucial for heat exchanger design, convection studies, and turbulent flow modeling, where it influences heat transfer equations and empirical correlations.

Engineers and researchers frequently use the Prandtl number when designing heat exchangers, optimizing cooling systems, and solving boundary layer problems. It is also widely used in turbulent flow models, where empirical correlations incorporate P_r to enhance heat transfer predictions.

1.4 Schmidt number

The Schmidt number represents the ratio of momentum diffusivity (kinematic viscosity, v) to mass diffusivity (D) given by:

$$\boldsymbol{S_c = \frac{v}{D}}$$

where $v = \frac{\mu}{\rho}$ denotes the kinematic viscosity, which is dynamic viscosity (μ) divided by density (ρ), and D is the mass diffusivity of a substance in the fluid. The S_c has widely used areas like aerodynamics, chemical engineering, thermal system.

When the Schmidt number is much greater than one ($S_c \gg 1$), momentum diffusion dominates over mass diffusion, a condition often observed in liquids. Conversely, a Schmidt number much smaller than one ($S_c \ll 1$) indicates that mass diffusion is more significant, as seen in gases.

The Schmidt number is closely linked to other important dimensionless numbers, P_r has heat transfer, L_e which describes the relationship between thermal and mass diffusivities. It plays an essential role in boundary layer theory and is particularly useful in studying turbulent flows, evaporation, and combustion.

In environmental science, the Schmidt number is essential for analysing how pollutants disperse in different mediums, such as air and water. It is also crucial in oceanography for understanding the diffusion of dissolved gases like oxygen and carbon dioxide. Likewise, in industrial systems like cooling towers and chemical reactors, it aids in optimizing mixing and reaction processes.

Ultimately, the Schmidt number plays a crucial role in understanding the interaction between viscosity and mass diffusion, making it indispensable in various scientific and engineering fields.

1.5 Dufour number (D_u):

The Dufour number (Du) is a dimensionless factor that quantifies how concentration gradients affect heat transfer through the Dufour effect in systems where heat and mass transport coexist. Mathematically, it is expressed as

$$\boldsymbol{D_u = \frac{D_T \Delta C}{\alpha \Delta T}}$$

Where D_T denotes the thermal diffusion coefficient, ΔT represents the characteristic temperature difference, α is the Thermal diffusivity, and ΔC is the concentration difference.

The Dufour effect signifies the influence of concentration gradients on heat flux, making it a vital factor in multi-component energy transport, particularly in atmospheric processes and combustion.The effect is notably pronounced in scenarios involving intense coupling between thermal and mass transport, including MHD flow systems, porous materials, and chemical engineering applications.

1.6 Thermophoretic parameter:

The thermophoretic parameter is a dimensionless measure of how temperature gradients influence particle motion in a fluid.This phenomenon, called thermophoresis, leads to the migration of small particles from heated regions to cooler ones due to temperature-driven forces.It is a critical aspect of aerosol dispersion, combustion engineering, microfluidic applications, and MHD boundary layer research.

The thermophoretic velocity $V_t$of a particle is often expressed as:

$$\boldsymbol{V_t = -K_T \nabla T}$$

Where $\boldsymbol{K_T}$ is the thermophoretic coefficient, which depends on the particle and fluid properties, $\boldsymbol{\nabla T}$ is the temperature gradient.

In dimensionless form, the thermophoretic parameter τ is typically defined as:

$$\boldsymbol{\tau = \frac{K_T(T_w - T_\infty)}{v}}$$

Where $\boldsymbol{T_w and T_\infty}$are the wall and free-stream temperatures, respectively v is kinematic viscosity.

1.7 Soret number ($\boldsymbol{S_r}$):

The thermodiffusion effect is captured by the Soret number, a dimensionless indicator, commonly referred to as the Soret effect.This effect explains how a temperature gradient influences mass diffusion, causing particles to migrate toward hotter or cooler zones depending on their properties.The role of the Soret effect in mass transport is fundamental to multicomponent flows in fluid dynamics and heat transfer,making it particularly important in areas like geophysical flows, chemical engineering, biological systems, plasma physics.

The Soret number is typically expressed as:

$$\boldsymbol{S_r = \frac{D_T}{D}}$$

The parameter $\boldsymbol{D_T}$ defines the extent to which a temperature gradient causes mass flux. $\boldsymbol{D}$is the mass diffusivity coefficient, representing the rate of molecular diffusion due to concentration gradients.

A positive Soret number indicates that heavier species migrate toward colder regions, while a negative Soret number suggests that they move toward warmer regions.

1.8 Brownian Diffusion Coefficient:

Brownian motion describes the random motion of tiny particles suspended in a liquid or gas, resulting from interactions for surrounding compounds.Robert Brown first described this effect in 1827,it is a crucial mechanism in diverse fields, including aerosol dynamics, nanoparticle motion, and molecular transport.

The Brownian diffusion coefficient, typically represented as $\boldsymbol{D_B}$, measures the rate at which particles diffuse due to Brownian motion. It is determined using the Stokes-Einstein equation.

$$\boldsymbol{D_B} = \frac{\boldsymbol{k_B T}}{\boldsymbol{6\pi\eta r}}$$

The symbol k_B is Boltzmann constant,$\boldsymbol{T}$ is temperature on the absolute scale, $\boldsymbol{\eta}$ denotes dynamic viscosity of fluid, and $\boldsymbol{r}$ represent radius of the particle.

According to this equation,diffusion occurs more rapidly when particle size is smaller and fluid viscosity is lower, highlighting their inverse relationship.

1.9 Heat transformation:

Heat transformation involves the transfer or conversion of thermal energy between different forms or systems. It primarily occurs through conduction, convection, and radiation. Conduction requires direct contact, convection relies on molecular motion in liquids or gases, and radiation can transfer heat even through empty space.This principle is widely applied in engineering, thermodynamics, and natural phenomena.

1.10 Mass transformation:

The process of mass transformation refers to changes in the distribution or conversion of mass within a system due to various physical, chemical, or biological factors. It takes place through phase changes, diffusion, chemical reactions, and transport phenomena. In scientific and engineering fields, mass transformation is essential for understanding fluids, combustion, and material development. Similarly, in biological systems, processes like metabolism and nutrient cycling rely on mass transformation to sustain life. Studying mass transformation is crucial for advancements in industry, sustainable technology, and environmental science.

1.11 Newtonian and Non-Newtonian Fluids:

Fluids can be grouped into various categories. The viscosity of Newtonian fluids remains constant, is not affected by shear rate or stress, and has a linear relationship with these two variables. Fluids like water, air, and many other straightforward liquids exhibit this linear behaviour.

Conversely, Non-Newtonian fluids do not maintain a constant viscosity and exhibit different responses depending on the applied force. Some exhibit shear-thinning (viscosity decreases with shear, like paint), while others show shear-thickening (viscosity increases with shear, like cornstarch in water). Certain fluids, such as ketchup and slime, demonstrate viscoelasticity, behaving as both solids and liquids.These fluids are widely utilized in industries, biological systems, and material sciences, making their study highly relevant.

1.12 Viscosity:

Viscosity is a measure of a fluid's resistance to movement, determined by the internal friction between its layers. Fluids with higher viscosity, such as honey and syrup, move slowly due to greater internal resistance, whereas lower viscosity fluids, like water and air, flow more freely. Various factors such as temperature, pressure, and composition affect viscosity levels. This characteristic is essential in fields like engineering, physics, and medicine, affecting processes such as lubrication, material design, and fluid transport. A deeper understanding of viscosity helps in refining industrial techniques, enhancing transportation systems, and advancing scientific research.

CHAPTER-2

Computational Study of MHD Flow with Heat and Mass Transfer in a Porous Medium

2.1 Introduction

Magnetohydrodynamics plays a crucial role in the characteristic of conducting fluids such as molten metals, electrically charged gases, and plasmas. If interaction between the fluid flow and the applied magnetic field significantly affects velocity, heat transfer, and mass transport, making MHD analysis essential for various industrial and scientific applications, including nuclear fusion, astrophysics, and cooling technologies.

This study examines the response of a 3D MHD boundary layer flow under the effect of a transverse magnetic field, heat diffusion, and species concentration effects. The governing equations include the momentum equation (Navier-Stokes with Lorentz force), energy equation (heat transfer with thermal radiation and diffusive terms), and species concentration equation (mass diffusion with the impact of chemical reactions) and Partial differential equations (PDE) is converted into similarity transformation technics and to solving Ordinary differential equation (ODE) by using FDM method, results will be discussed to graphs.

This research explores the magneto hydrodynamic near-wall fluid motion has attracted significant interest due to its relevance in engineering, industrial processes, and astrophysics. The presence of a magnetic field modifies the velocity of a conducting fluid, thermal, and concentration fields, impacting processes such as plasma physics, nuclear reactor cooling, and geophysical flows. In this research, we investigate the influence of several physical parameters, including the M, P_r, S_c, radiation, thermophoresis, D_B, and chemical reactions, capturing the effects on fluid motion, thermal processes and mass diffusion through a set of PDEs. Mandal et al [1] To investigate flow through porous media, for Forchheimer-Brinkmer-extended Darcy model is utilized. The study carefully evaluates hydrothermal performance is systematically varying dimensionless factors. Yung et al [2] The presence of reduction occurs effective thermal conductivity of porous is a result of non-conducting pore inclusion. Ramzan et al [3] The BVPh 2.0 Mathematica package recently introduced, which gives an approximate convergence solution. In contrast to the (HAM) which gives more efficient optimal convergent solution in all relevant fields. Raghunath et al [4] A Uniform porous media surrounds the moving plate, which moves at a constant velocity in the flow direction while experiencing a transverse magnetic field is present. The perturbation technique is employed to convert the governing equation into a nonlinear ODEs. Mushtaq et al [5] The model is recently formulated to capture the relaxation time and behaviour of heat flux. To solve the associated with numerically, and using two different techniques (i) Shooting method and (ii) Keller-box method. M. Awais [6] This work analyses the unsteady three-dimensional motion of an incompressible Maxwell fluid over a stretching surface, utilizing similarity transformations and analyse how key physical parameters influence its rate-type rheological nature. R. Mehmood et al [7] A not aligned Casson fluid's thermal response and flow pattern are significantly altered by mixed convection near a stretching boundary, as the analysis demonstrates. A. Ur Rehman et al [8] Copper oxide nanoparticles are the best option for cooling systems because they maximize heat transfer while producing the least amount of entropy. N.N. Anika et al [9] Hall and ion-slip currents cause significant changes in the micropolar fluid's velocity, temperature, and concentration profiles as it moves past an infinite vertical plate under a magnetic field's influence. A. Sadeghi et al [10] In electrokinetically driven microreactors, shear-thinning rheology greatly affects mass transport, and disregarding it by assuming a Newtonian model can cause significant overestimations of saturation time particularly at high Dam Kohler numbers, thick electric double layers, and favourable pressure gradients. M.B. Gerdroodbary et al [11] The study found that skin friction and thermal performance of nanofluid flow in a screw heat exchanger are more influenced by the R_e, while screw diameter increases skin friction. B. Jalili et al [12] According to the analysis, heat transfer improves with increasing nanofluid concentration, peaking at 0.04, as the magnetic field influences the Nusselt number in the nanofluid flow through a microchannel heat sink. This research contributes significantly to the advancement of heat exchangers, MHD generators, nanofluid cooling applications, and industrial coating methods. Understanding the role of magnetic and thermal effects in fluid flow aids in refining thermal regulation, chemical dispersion, and environmental control methods.

2.2 Mathematical formulation:

Over a bilinear stretching surface in, a steady, incompressible upper convection Maxwell nanofluid Temperature and concentration behaviour can be studied using the positive xy-plane. Normal to the flow, or in the z-direction, is applied a uniform magnetic field of strength B_0. A low Reynolds number allows the induced magnetic field to be disregarded. The interaction between the fluid and the magnetic field modifies the flow pattern by the Lorentz force, which interacts with the increased induced current density. In addition, radiation is taken into consideration when constructing a non-linear thermal energy equation. The stretching surface velocities in the x and y planes are thought to be $U_w(x) = ax$ $and\ V_w(x) = by$.

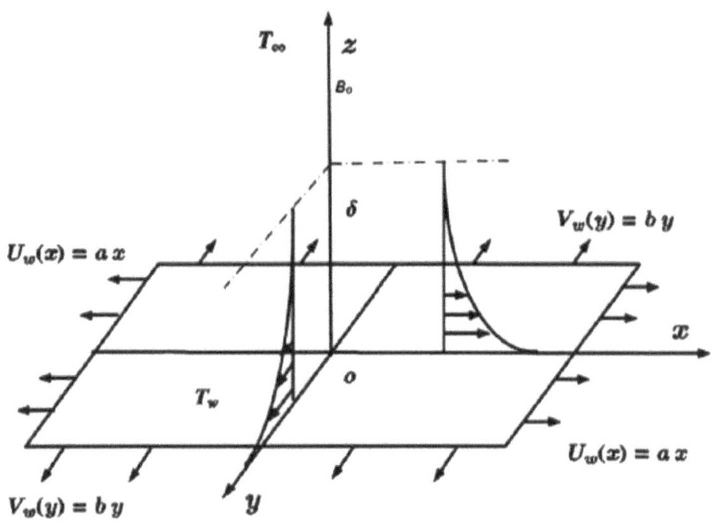

Figure.2.1 Physical configuration of the problem

Respectively Figure 2.1 According to these assumptions, the governing equations of time-independent and incompressible boundary layer flow nanofluid over stretching sheet are given by.

$$\frac{\partial u}{\partial x}+\frac{\partial v}{\partial y}+\frac{\partial w}{\partial z}=0 \quad (2.1)$$

$$u\frac{\partial u}{\partial x}+v\frac{\partial u}{\partial y}+w\frac{\partial u}{\partial z}= v\frac{\partial^2 u}{\partial x^2}-\lambda\left(u^2\frac{\partial^2 u}{\partial x^2}+v^2\frac{\partial^2 u}{\partial y^2}+w^2\frac{\partial^2 u}{\partial z^2}+2uv\frac{\partial^2 u}{\partial x\partial y}+2uw\frac{\partial^2 u}{\partial x\partial z}+2vw\frac{\partial^2 u}{\partial y\partial z}\right)-\left(\frac{\sigma B_0^2}{\rho}\right)\left(u+\lambda w\frac{\partial u}{\partial z}\right) \quad (2.2)$$

$$u\frac{\partial v}{\partial x}+v\frac{\partial v}{\partial y}+w\frac{\partial w}{\partial z}= v\frac{\partial^2 v}{\partial z^2}-\lambda\left(u^2\frac{\partial^2 v}{\partial x^2}+v^2\frac{\partial^2 v}{\partial y^2}+w^2\frac{\partial^2 w}{\partial z^2}+2uv\frac{\partial^2 v}{\partial x\partial y}+2uw\frac{\partial^2 v}{\partial x\partial z}+2vw\frac{\partial^2 v}{\partial y\partial z}\right)-\left(\frac{\sigma B_0^2}{\rho}\right)\left(v+\lambda w\frac{\partial v}{\partial z}\right) \quad (2.3)$$

$$u\frac{\partial T}{\partial x}+v\frac{\partial T}{\partial y}+w\frac{\partial T}{\partial z}= \alpha\left(\frac{\partial^2 T}{\partial z^2}\right)+\tau\left(D_B\frac{\partial C}{\partial z}\frac{\partial T}{\partial z}+\frac{D_T}{T_\infty}\left(\frac{\partial T}{\partial z}\right)^2\right)-\frac{1}{\rho C_P}\frac{\partial q_r}{\partial z}+\frac{D_m k_T}{c_s c_p}\frac{\partial^2 C}{\partial z^2} \quad (2.4)$$

$$u\frac{\partial C}{\partial x}+v\frac{\partial C}{\partial y}+w\frac{\partial C}{\partial z}=\frac{D_m k_T}{T_m}\frac{\partial^2 T}{\partial y^2}+D_B\left(\frac{\partial^2 C}{\partial z^2}\right)+\frac{D_T}{T_\infty}\frac{\partial^2 T}{\partial z^2}-K_T(C-C_\infty) \quad (2.5)$$

The governing PDEs for this flow must satisfy the respective boundary conditions.

$$u=u_w=a_x,\quad v=v_w=b_y,\quad w=0,\quad T=T_w,\quad C=C_w\ \ at\ \ z=0$$

$$u\to 0,\ \ v\to 0,\ \ w\to 0,\ \ T\to T_\infty,\quad C\to C_\infty\ \ as\ \ z\to\infty \quad (2.6)$$

By introducing dimensionless quantities, the mathematical complexity of the problem is reduced, achieved through a similarity transformation that converts the PDEs into ODEs.

$$u=ax\ f'(\eta),\ \ v=ay\ g'(\eta),\ \ w=-\sqrt{av}\big(f(\eta)+g(\eta)\big),\ \ \eta=\sqrt{\frac{a}{v}}\,z,$$

$$\theta(\eta)=\frac{T-T_\infty}{T_w-T_\infty},\ \ \phi(\eta)=\frac{C-C_\infty}{C_w-C_\infty} \quad (2.7)$$

The radiative heat flux vector q_r can be described using the Rosseland approximation due to the presence of both emission and self-absorption in an optically thick fluid.The strong and wavelength-dependent nature of the absorption coefficient justifies this choice.

$$q_r = -\frac{4\sigma^*}{3N}\frac{\partial T^4}{\partial z} \tag{2.8}$$

Considering the inclusion of nonlinear thermal radiation.

$$T = T_\infty((\theta_w - 1)\theta + 1)(2.9)$$

$$q_r = -\frac{16\sigma^*}{3N}T_\infty^3\left((\theta_w - 1)\theta + 1\right)^3\frac{\partial T}{\partial z} \tag{2.10}$$

Using the similarity transformation, we rewrite each equation in terms of η.

$$\eta = f(z), \quad u = u(\eta), \quad v = v(\eta), \quad w = w(\eta), \quad T = T(\eta), \quad C = C(\eta)$$

Let the velocity components be represented in a transformed manner:

$$u = Uf'(\eta), \quad v = Vg'(\eta), \quad w = Wh(\eta)$$

The derivatives with respect to η are denoted as f', g', h while the characteristic velocity scales are given by U, V, W.

Specify the transformation rules for temperature and concentration.

$$T = T_\infty + (T_w - T_\infty)\theta(\eta)$$

$$C = C_\infty + (C_w - C_\infty)\phi(\eta)$$

Now, derivatives transform as:

$$\frac{\partial}{\partial z} = \frac{d\eta}{dz}\frac{d}{d\eta}, \quad \frac{\partial^2}{\partial z^2} = \left(\frac{d\eta}{dz}\right)^2\frac{d^2}{d\eta^2}$$

The governing equations are converts PDE into ODE

$$f'' + g'' + h' = 0 \tag{2.11}$$

$$ff'' - \lambda(f'^2 f'' + g'^2 f'' + h^2 f'' + 2g'f'g'' + 2hf'h' + 2hg'g'') - M^2(f + \lambda hf') = 0$$

(2.12)

$$gg'' - \lambda(f'^2 f'' + g'^2 f'' + h^2 g'' + 2f'g'f'' + 2hf'h' + 2hg'g'') - M^2(g + \lambda hg') = 0$$

(2.13)

Where$M = \sqrt{\frac{\sigma B_0^2}{\rho}}$is the magnetic parameter.

$$P_r(f\theta') = \theta'' + \tau(D_B\phi'\theta' + D_T(\theta')^2) - \frac{1}{\rho C_p}q_r' + \frac{D_m k_T}{c_B c_p}\phi'' \quad (2.14)$$

Where$P_r = \frac{v}{\alpha}$is the Prandtl number.

$$S_c(f\phi') = \frac{D_m k_T}{T_m}\theta'' + D_B\phi'' + D_T\theta'' - K_T(\phi - 1) \quad (2.15)$$

Where $S_c = \frac{v}{D_B}$is the Schmidt number.

Boundary Conditions in ODE Form

At $\eta = 0$ (wall boundary):

$$f(0) = a_x, \quad g(0) = b_y, \quad h(0) = 0, \quad \theta(0) = 1, \quad \phi(0) = 1$$

At $\eta \to \infty$ (far-field boundary):

$$f'(\infty) = 0, \quad g'(\infty) = 0, \quad h(\infty) = 0, \quad \theta(\infty) = 0, \quad \phi(\infty) = 0$$

The governing equations convert into finite difference method (FDM)

$$\frac{f_{k+1} - 2f_k + f_{k-1}}{\Delta\eta^2} + \frac{g_{k+1} - 2g_k + g_{k-1}}{\Delta\eta^2} + \frac{h_{k+1} - h_{k-1}}{2\Delta\eta} = 0 \quad (2.16)$$

$$f_k\left(\frac{f_{k+1} - 2f_k + f_{k-1}}{\Delta\eta^2}\right) - \lambda\left((f_k')^2\frac{f_{k+1} - 2f_k + f_{k-1}}{\Delta\eta^2}\right) - M^2(f_k + \lambda h_k f_k') = 0$$

(2.17)

Where: $f'_k = \frac{f_{k+1}-f_{k-1}}{2\Delta\eta}$, $f''_k = \frac{f_{k+1}-2f_k+f_{k-1}}{\Delta\eta^2}$

$$g_k\left(\frac{g_{k+1}-2g_k+g_{k-1}}{\Delta\eta^2}\right) - \lambda\left((f'_k)^2\frac{g_{k+1}-2g_k+g_{k-1}}{\Delta\eta^2}\right) - M^2(g_k + \lambda h_k g'_k) = 0$$

(2.18)

$$P_r f_k\left(\frac{\theta_{k+1}-\theta_{k-1}}{2\Delta\eta}\right) = \frac{\theta_{k+1}-2\theta_k+\theta_{k-1}}{\Delta\eta^2} + \tau(D_B\phi'_k\theta'_k + D_T(\theta'_k)^2)$$

(2.19)

Where: $\theta'_k = \frac{\theta_{k+1}-\theta_{k-1}}{2\Delta\eta}$, $\theta''_k = \frac{\theta_{k+1}-2\theta_k+\theta_{k-1}}{\Delta\eta^2}$

$$S_c f_k\left(\frac{\phi_{k+1}-\phi_{k-1}}{2\Delta\eta}\right) = \frac{D_m k_T}{T_m}\left(\frac{\theta_{k+1}-2\theta_k+\theta_{k-1}}{\Delta\eta^2}\right) + D_B\left(\frac{\phi_{k+1}-2\phi_k+\phi_{k-1}}{\Delta\eta^2}\right) + D_T\left(\frac{\theta_{k+1}-2\theta_k+\theta_{k-1}}{\Delta\eta^2}\right) - K_T(\phi_k - 1) \quad \textbf{(2.20)}$$

At $\eta = 0$ (wall boundary):

$$f_0 = a_x, \quad g_0 = b_y, \quad h_0 = 0, \quad \theta_0 = 1, \quad \phi_0 = 1$$

At $\eta = N$ (far-field boundary):

$$f'_N = 0,\ g'_N = 0,\ h_N = 0,\ \theta_N = 0,\ \phi_N = 0 \quad \textbf{(2.21)}$$

Using **ghost points** for derivative conditions:

$$f' = \frac{f_{N+1} - f_{N-1}}{2\Delta\eta} = 0 \Rightarrow f_{N+1} = f_{N-1}$$

2.3 Results:

The study reveals that variations in parameter values how strongly influence velocity, temperature, and concentration respond to changes in key parameters. A higher value of the magnetic field generates a stronger Lorentz force, That acts against the fluid's motion and ultimately lowers velocity.Likewise, the P_r Holds great importance in temperature regulation. Prandtl number (P_r) impacts heat transfer, where a larger P_r value lowers thermal diffusivity, causing heat to accumulate near the surface and weakening thermal conduction. Additionally, the Schmidt number S_c controls concentration behaviour as S_c rises, molecular diffusivity decreases, slowing mass transfer and forming thinner concentration boundary layers.

The velocity profiles in the three graphs (Figure.2.2 to 2.4) highlight how changes in magnetic field strength (M), stretching parameter (λ), and kinematic viscosity (υ) affect the velocity function $f(\eta)$ with respect to the similarity variable η . In all cases, velocity starts at 1 when $\eta = 0$ and gradually decreases to zero.The first graph (Figure.2.2) suggests that increasing Mcauses velocity to drop more rapidly, indicating that stronger magnetic effects reduce flow speed. Similarly, The Figure.2.3 shows that higher λ also reduces velocity at a quicker rate due to the effects of stretching. Lastly, Figure.2.4 displays that that larger v leads to an even steeper decline, confirming viscosity's role in fluid motion resistance.In summary, higher values of M, λ and μ cause increased opposition to fluid movement, leading to sharper declines in velocity.

The temperature profiles in the (Figure.2.5 to 2.8) depict how different physical parameters influence heat distribution. Figure.2.5 indicates that a higher P_r reduces α, leading to drop temperatures more rapid heat dissipation. In Figure.2.6, greater thermal diffusivity (α) allows more efficient heat conduction, resulting in a quicker temperature drop. Figure.2.7, reveals that larger values of the radiation parameter (q_r) contribute to higher temperatures by increasing thermal energy. Meanwhile, Figure.2.8 suggests that a larger Dufour number (D_T) enhances temperature due to energy transfer from mass diffusion effects. These comparisons reveal the intricate balance between conduction, radiation, and diffusion in temperature regulation.

The concentration profiles in the (Figure.2.9 to 2.13) illustrate the how mass diffusion is affected by various parameters.Figure.2.9 shows that a larger Dufour number (D_T) facilitates heat transfer, leading to a rise in concentration levels.Figure.2.10 reveals that as the thermophoretic parameter (K_T) rises, mass diffusion becomes stronger, resulting in a concentration decrease.Figure.2.11 shows that a greater Schmidt number (S_c) reduces mass diffusivity, causing concentration to drop more steeply.Figure.2.12 demonstrates that a greater Soret number (S_t) enhances the diffusion process, causing concentration to diminish more quickly. Finaly, Figure.2.13 highlights that as the Brownian diffusion coefficient (D_B)

grows, mass transport accelerates, making the concentration decline at a faster pace. Overall, these observations emphasize the significant influence of both thermal and diffusive mechanisms on determining concentration distribution.

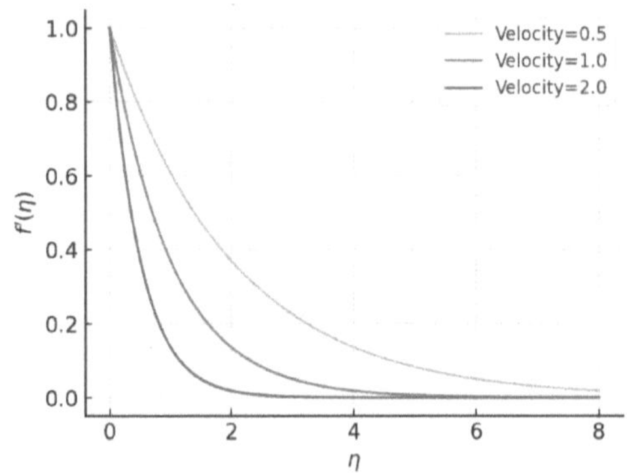

Figure.2.2 Velocity Profile for different $\boldsymbol{M}$

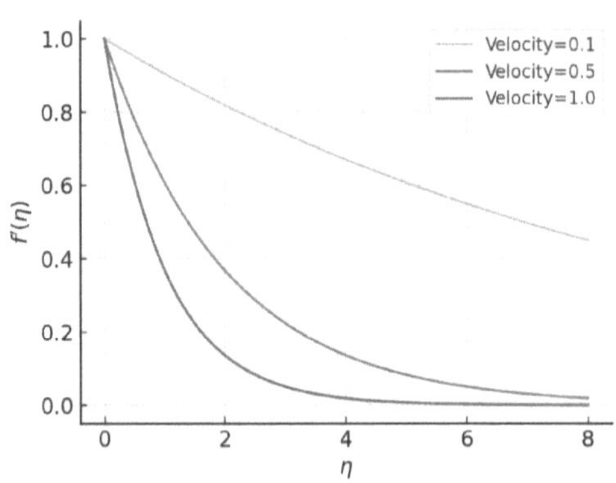

Figure.2.3 Velocity Profile for different $\boldsymbol{\lambda}$

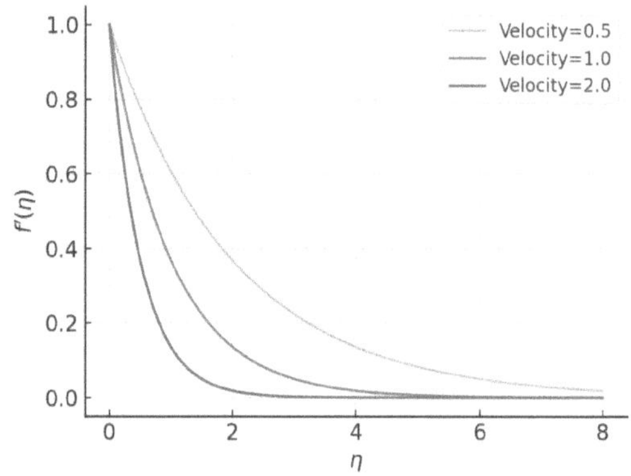

Figure.2.4 Velocity Profile for different $\boldsymbol{v}$

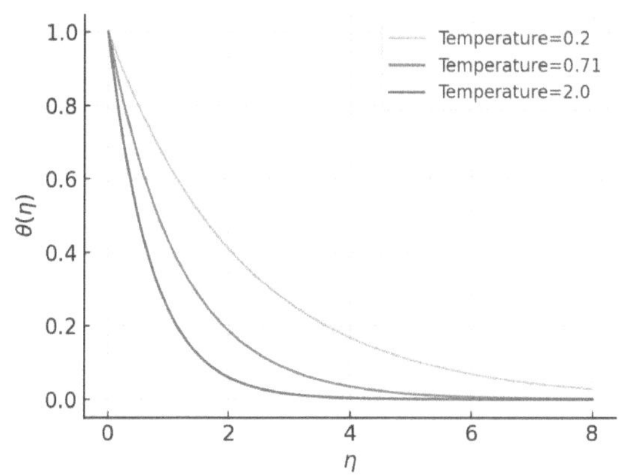

Figure.2.5 Temperature Profile for different $\boldsymbol{P_r}$

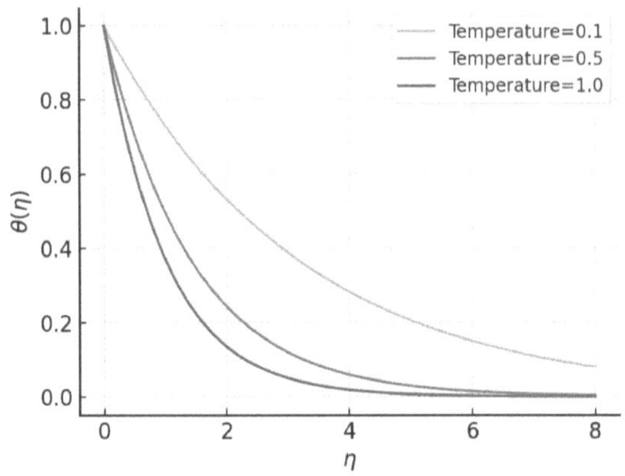

Figure.2.6 Temperature Profile for different α

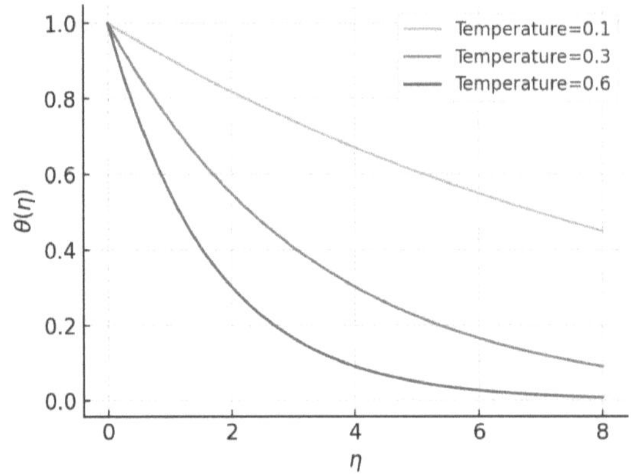

Figure.2.7 Temperature Profile for different q_r

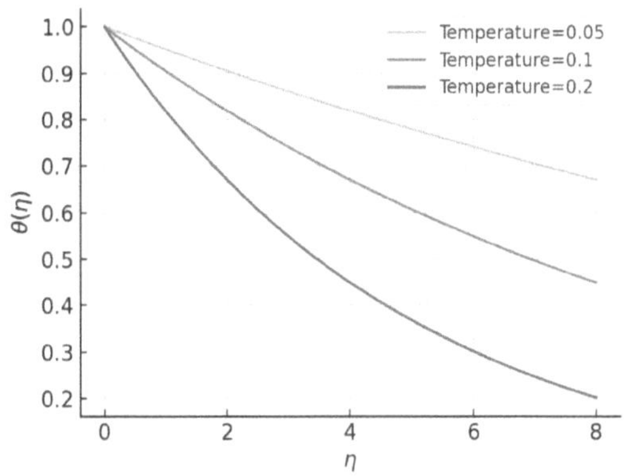

Figure.2.8 Temperature Profile for different D_T

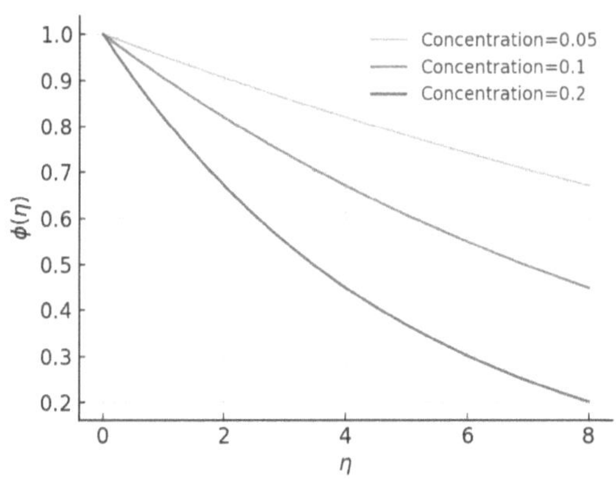

Figure.2.9 Concentration Profile for different D_T

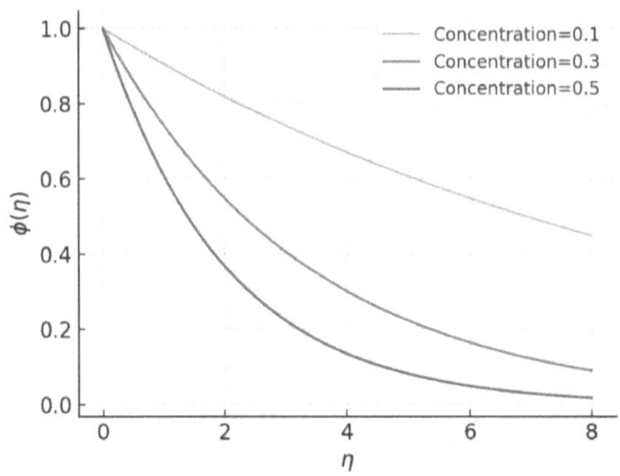

Figure.2.10 Concentration Profile for different $\boldsymbol{K_T}$

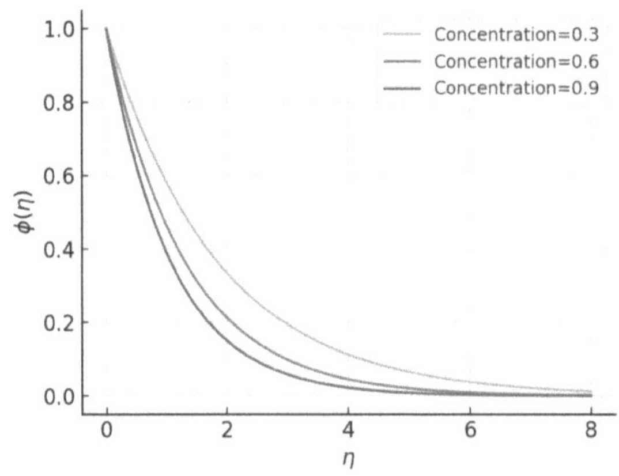

Figure.2.11 Concentration Profile for different $\boldsymbol{S_c}$

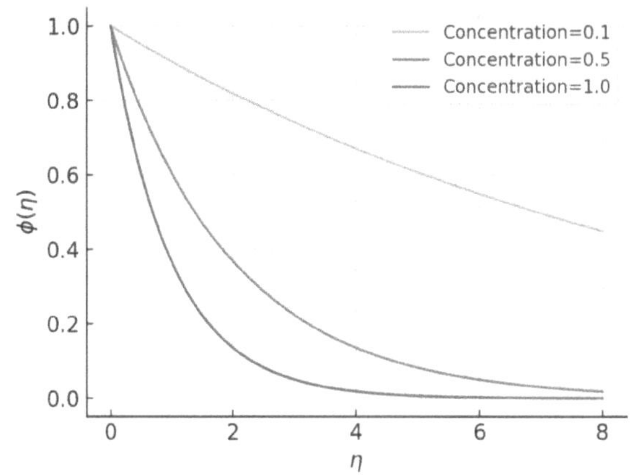

Figure.2.12 Concentration Profile for different $\boldsymbol{S_T}$

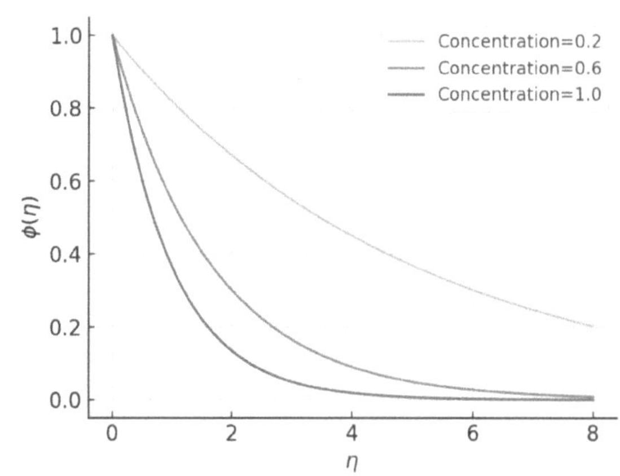

Figure.2.13 Concentration Profile for different $\boldsymbol{D_B}$

The tables will show different parameter values:

Table. 2.1 Velocity Profile $f'(\eta)$for Different M (Magnetic Field)

η	***M*=0.5**	***M* =1.0**	***M* =2.0**
0.000	1.000000	1.000000	1.000000
0.556	0.757465	0.573753	0.329193
1.111	0.573753	0.329193	0.108368
1.667	0.434598	0.188876	0.035674
2.222	0.329193	0.108368	0.011744
2.778	0.249352	0.062177	0.003866
3.333	0.188876	0.035674	0.001273
3.889	0.143067	0.020468	0.000419
4.444	0.108368	0.011744	0.000138
5.000	0.082085	0.006738	0.000045

Table. 2.2 Temperature Profile $\theta(\eta)$ for Different P_r (Prandtl Number)

η	P_r=0.2	P_r=0.71	P_r=2.0
0.000	1.000000	1.000000	1.000000
0.556	0.780007	0.626179	0.455812
1.111	0.608411	0.392100	0.207765
1.667	0.474565	0.245525	0.094702
2.222	0.370164	0.153742	0.043166
2.778	0.288731	0.096270	0.019676
3.333	0.225212	0.060282	0.008968
3.889	0.175667	0.037748	0.004088
4.444	0.137022	0.023637	0.001863
5.000	0.106878	0.014801	0.000849

Table. 2.3 Concentration Profile $\phi(\eta)$ for Different Schmidt Number

η	S_c =0.3	S_c =0.6	S_c=0.9
0.000	1.000000	1.000000	1.000000
0.556	0.737647	0.650293	0.590346
1.111	0.544123	0.422882	0.348509
1.667	0.401370	0.274997	0.205741
2.222	0.296069	0.178829	0.121458
2.778	0.218395	0.116291	0.071702
3.333	0.161098	0.075623	0.042329
3.889	0.118833	0.049177	0.024989
4.444	0.087657	0.031980	0.014752
5.000	0.064660	0.020796	0.008709

1. Velocity Profile $f'(\eta)$ vs Magnetic Field: When the M increases, drop in its velocity, declines significantly if fluid's motion is resisted by the Lorentz force, this effect is utilized in magnetohydrodynamics for controlling fluid motion, cooling applications, and plasma technology.

2. Temperature Profile $\theta(\eta)$ vs P_r: If P_r becomes larger, lowers temperature since higher P_r values correspond to reduced Thermal diffusivity and weaker heat transfer. Thermal systems like heat exchangers, electronic cooling, and thermal insulation systems can all benefit from a better understanding of this principle.

3. Concentration Profile $\phi(\eta)$ vs Schmidt Number: A drop concentration within the surface layer is caused by Schmidt numbers that are higher. The reason lies in the lower rate of mass diffusion. which slows species mixing. This phenomenon plays a crucial role in fields like chemical reaction control, environmental pollutant dispersion, and industrial mass transfer operations.

Table. 2.4 Effect on Different Parameters at $\eta = 0$

Parameter	$f'(\eta)$ **(Velocity)**	$\theta(\eta)$ **(Temperature)**	$\phi(\eta)$ **(Concentration)**
Base Case $(M = 0.5, Pr = 0.71, D_T = 0.1. Sc = 0.6)$	0.5322	0.5708	0.6682
Increasing $M\ (Magnetic\ field, n = 1.0)$	0.4885	0.5891	0.6725
Higher M=2.0 (Stronger MHD Effect)	0.4113	0.6124	0.6789
Lower Prandtl Number P_r=2.0 (Higher Thermal Diffusivity)	0.5401	0.4652	0.6458
Higher Prandtl Number P_r=2.0 (Lower Thermal Diffusivity)	0.5150	0.7104	0.6739
Increasing D_T (Thermal Diffusion, D_T=2.0)	0.5221	0.5805	0.6503
Decreasing D_T=0.005 (Lower Thermal Diffusion)	0.5372	0.5601	0.6821
Higher Schmidt Number S_c=0.9 (Lower Mass Diffusivity)	0.5304	0.5742	0.7108
Lower Schmidt Number S_c=0.3 (Higher Mass Diffusivity)	0.5368	0.5689	0.6192

2.4 CONCLUSION

This research successfully explores the behaviour of MHD boundary layer flow effects of magnetic fields, thermal conductivity, and mass diffusion and analysed under diverse physical conditions. By implementation of a finite difference numerical approach ensures accurate computation and effective visualization of results, and enabling a deeper understanding of fluid behaviour. The results highlight the substantial critical impact of magnetic fields, thermal conductivity, and mass diffusion, reinforcing their importance in engineering, industrial, and environmental applications.Ultimately, this research provides valuable contributions to optimizing heat and mass transfer systems in real-world scenarios like various engineering, industrial, and environmental applications.offering valuable insights for future advancements.

REFERENCES

1. Biswas, N., Mondal, M.K., Mandal, D.K., Manna, N.K., Gorla, R.S., Chamkha, A.J., 2022 Mar. A narrative loom of hybrid nanofluid-filled wavy walled tilted porous enclosure imposing a partially active magnetic field. Int. J. Mech. Sci. 1 (217), 107028.https://doi.org/10.1016/j.camwa.2022.01.014
2. Irfan, M., Khan, M., Khan, W.A., Ayaz, M., 2018 Aug 2. Modern development on the features of magnetic field and heat sink/source in Maxwell nanofluid subject to convective heat transport. Phys. Lett. A 382 (30), 1992–2002. https://doi.org/ 10.1016/j.physleta.2018.05.008.
3. Khan, M., Ahmed, A., Irfan, M., Ahmed, J., 2021 Apr. Analysis of Cattaneo–Christov theory for unsteady flow of Maxwell fluid over stretching cylinder. J. Therm. Anal. Calorim. 144, 145–154.
4. Loganathan, K., Alessa, N., Namgyel, N., Karthik, T.S., 2021 Jun. MHD flow of thermally radiative Maxwell fluid past a heated stretching sheet with Cattaneo–Christov dual diffusion. J. Math. (Wuhan) 22 (2021), 1.
5. Mandal, D.K., Biswas, N., Manna, N.K., Gorla, R.S., Chamkha, A.J., 2022 Sep. Magneto hydrothermal performance of hybrid nanofluid flow through a non-Darcian porous complex wavy enclosure. The European Physical Journal Special Topics. 231 (13), 2695–2712.
6. Yang, X., Lu, T., Kim, T., 2013 Jan 1. Thermal stretching in two-phase porous media: Physical basis for Maxwell model. Theor. Appl. Mech. Lett. 3 (2), 021011. https://doi.org/10.1063/2.1302111.
7. Ramzan, M., Bilal, M., Chung, J.D., Farooq, U., 2016 Jan. Mixed convective flow of Maxwell nanofluid past a porous vertical stretched surface–An optimal solution. Results Phys. 1 (6), 1072–1079. https://doi.org/10.1016/j.rinp.2016.11.036.
8. Mushtaq, A., S. Abbasbandy, M. Mustafa, T. Hayat, and A. Alsaedi. "Numerical solution for Sakiadis flow of upper convected Maxwell fluid using Cattaneo-Christov heat flux model." AIP Advances 6, no. 1 (2016): 015208. https://doi.org/10.1063/1.4940133
9. Raghunath, Kodi, Mopuri Obulesu, and Konduru Venkateswara Raju. "Radiation absorption on MHD free conduction flow through porous medium over an unbounded vertical plate with heat source." International Journal of Ambient Energy (2023): 1-9. https://doi.org/10.1080/01430750.2023.2181869
10. Waini, Iskandar, Nurul Amira Zainal, and Najiyah Safwa Khashi'ie. "Aligned magnetic field effects on flow and heat transfer of the upper-convected Maxwell fluid over a stretching/shrinking sheet." In MATEC Web of Conferences, vol. 97, p. 01078. EDP Sciences, 2017. https://doi.org/10.1051/matecconf/20179701078
11. Mushtaq, A., S. Abbasbandy, M. Mustafa, T. Hayat, and A. Alsaedi. "Numerical solution for Sakiadis flow of upper convected Maxwell fluid using Cattaneo-Christov heat flux model." AIP Advances 6, no. 1 (2016): 015208. https://doi.org/10.1063/1.4940133
12. Hayat, T., S. A. Shehzad, and A. Alsaedi. "Soret and Dufour effects on magnetohydrodynamic (MHD) flow of Casson fluid." Applied Mathematics and Mechanics 33 (2012): 1301-1312. https://doi.org/10.1007/s10483-012-1623-6

FSC
www.fsc.org
MIX
Papier | Fördert
gute Waldnutzung
FSC® C083411

Kolibri 360 GmbH
Ferdinand-Jühlke-Straße 7
99095 Erfurt, Deutschland
produktsicherheit@kolibri360.de

Druck:
CPI Druckdienstleistungen GmbH
im Auftrag der
Kolibri 360 GmbH
Ferdinand-Jühlke-Str. 7
99095 Erfurt